Frontiers in Mathematics

Michael Huber

Flag-transitive
Steiner
Designs

Birkhäuser Verlag
Basel · Boston · Berlin

Author:
Michael Huber
Technical University Berlin
Institute of Mathematics
Straße des 17. Juni 136
10623 Berlin
Germany
e-mail: mhuber@math.tu-berlin.de

2000 Mathematical Subject Classification: Primary 51E10; Secondary 05B05, 20B25

Library of Congress Control Number: 2008939514

Bibliographic information published by Die Deutsche Bibliothek
Die Deutsche Bibliothek lists this publication in the Deutsche Nationalbibliografie;
detailed bibliographic data is available in the Internet at <http://dnb.ddb.de>.

ISBN 978-3-0346-0001-9 Birkhäuser Verlag AG, Basel · Boston · Berlin

© 2009 Birkhäuser Verlag AG
Basel · Boston · Berlin
P.O. Box 133, CH-4010 Basel, Switzerland
Part of Springer Science+Business Media
Cover design: Birgit Blohmann, Zürich, Switzerland
Printed on acid-free paper produced from chlorine-free pulp. TCF ∞
Printed in Germany

ISBN 978-3-0346-0001-9   e-ISBN 978-3-0346-0002-6

9 8 7 6 5 4 3 2 1   www.birkhauser.ch

Contents

Preface

The characterization of combinatorial or geometric structures in terms of their groups of automorphisms has attracted considerable interest in the last decades and is now commonly viewed as a natural generalization of Felix Klein's Erlangen program (1872). In addition, especially for finite structures, important applications to practical topics such as design theory, coding theory and cryptography have made the field even more attractive.

The subject matter of this research monograph is the study and classification of flag-transitive Steiner designs, that is, combinatorial t-$(v, k, 1)$ designs which admit a group of automorphisms acting transitively on incident point-block pairs. As a consequence of the classification of the finite simple groups, it has been possible in recent years to characterize Steiner t-designs, mainly for $t = 2$, admitting groups of automorphisms with sufficiently strong symmetry properties. For Steiner 2-designs, arguably the most general results have been the classification of all point 2-transitive Steiner 2-designs in 1985 by W. M. Kantor, and the almost complete determination of all flag-transitive Steiner 2-designs announced in 1990 by F. Buekenhout, A. Delandtsheer, J. Doyen, P. B. Kleidman, M. W. Liebeck, and J. Saxl.

However, despite the classification of the finite simple groups, for Steiner t-designs with $t > 2$ most of the characterizations of these types have remained long-standing challenging problems. Specifically, the determination of all flag-transitive Steiner t-designs with $3 \leq t \leq 6$ has been of particular interest and object of research for more than 40 years.

The main part of this monograph is devoted to the complete classification of all flag-transitive Steiner t-designs for each of the remaining parameters $t = 3, 4, 5, 6$. The obtained results generalize theorems of Jacques Tits (1964) and Heinz Lüneburg (1965). The primary objects that are characterized are the Mathieu-Witt designs associated with the five sporadic simple Mathieu groups; thus the results are also important for a future unified geometric theory of the sporadic simple groups. The proofs rely on the classification of the finite 2-transitive permutation groups, which itself depends on the finite simple group classification. Along with group theory, the proofs also involve incidence geometric, combinatorial and number theoretical arguments. Especially for the latter, the study of

Diophantine equations, in particular Thue-Mahler and generalized Ramanujan-Nagell equations, turns out to be helpful for crucial parts of the proofs. The main results have been published recently [59, 60, 61, 62, 63], and are presented in this treatment in a sufficiently self-contained and unified manner. Moreover, a broad introduction to the topic of flag-transitive Steiner designs is provided, along with illustrative examples.

Here is a brief chapter-by-chapter description of the contents; a more detailed one may be found at the beginning of each chapter.

Chapters 1–3 are of expository nature; Chapter 1 gives an introduction to the theory of incidence structures and combinatorial designs; Chapter 2 is on permutation groups and group actions, in particular the classification of the finite doubly transitive permutation groups is stated; Chapter 3 assembles number-theoretical tools like Zsigmondy's theorem on primitive prime divisors and related issues. The advanced reader may skip these first three chapters.

In Chapter 4, we start to look at Steiner designs which admit a group of automorphisms with sufficiently strong symmetry properties. One of the reasons for this consideration of highly symmetric designs is a general view that, while the existence of combinatorial objects is of interest, they are even more fascinating when they have a rich group of symmetries. Various examples are illustrated, most of them arising from finite geometries. Among the highly symmetric properties of designs, flag-transitivity is certainly a particularly important and natural one, and hence will be of further central consideration. In particular, as the starting point of our examination of all flag-transitive Steiner designs, we derive the following result: For any non-trivial t-design $\mathcal{D}$ with $t \geq 3$, the flag-transitivity of a group $G \leq \mathrm{Aut}(\mathcal{D})$ of automorphisms of $\mathcal{D}$ always implies its doubly transitivity on the points of $\mathcal{D}$. The proof involves Block's Lemma, a well-known result which is also treated in detail in this chapter. In the next chapter, Chapter 5, the complete determination of all flag-transitive Steiner t-designs with $t \geq 3$ is stated. Moreover, a census of some of the most general results on highly symmetric Steiner t-designs is given.

The rest of the book is dedicated to the complete classification of all flag-transitive Steiner t-designs with $3 \leq t \leq 6$: First, in Chapter 6 we classify all flag-transitive Steiner quadruple systems, i.e., Steiner 3-designs with block size 4. The key ideas of the proof are presented at a level suitable for beginning graduate students. In a more rigorously mathematical way, these results are extended in Chapter 7 to arbitrary Steiner 3-designs. Chapter 8 deals with the determination of all flag-transitive Steiner 4-designs. We present the classification of all flag-transitive Steiner 5-designs in Chapter 9 and prove finally in Chapter 10 that there are no non-trivial flag-transitive Steiner 6-designs.

This book provides the first full discussion of flag-transitive Steiner designs, a central part of the study of highly symmetric combinatorial configurations at the interface of several mathematical disciplines. It is addressed to graduate students in mathematics or computer science with some familiarity with combinatorics and basic group theory as well as to established researchers in design theory, finite or incidence geometry, coding theory, cryptography, algebraic combinatorics, and more generally, discrete mathematics.

I want to thank Francis Buekenhout, Peter Cameron, Reinhard Laue, Alex Lubotzky, Bill Kantor, Richard Stanley, and Günter Ziegler, among others, for helpful conversations, their encouragement and support, as well as the Deutsche Forschungsgemeinschaft (DFG), Heisenberg-Programme, Institut für Mathematik of the Technische Universität Berlin, Mathematisches Institut of the Universität Tübingen. I am especially grateful to Christoph Hering for his constant support. Most of all, I thank my family: my wife Susanne and our two daughters Lynn-Scharon and Sheila-Ann.

Berlin,   *Michael Huber*
September 2008

Chapter 1

Incidence Structures and Steiner Designs

1.1 Introduction

Combinatorial design theory is a fascinating subject of considerable interest in discrete mathematics and computer science, amongst others. It deals with a crucial problem of combinatorial theory, namely, that of arranging objects into patterns according to specified rules.

We give in this chapter a brief introduction to the topic, with emphasis on Steiner designs. For a more general treatment, the reader is referred to [8, 12, 24, 34, 51, 70, 89, 116]. In particular, [8, 34] provide excellent encyclopedic accounts of key results.

From the many connections of design theory to other fields, we mention in our context especially its links to finite and incidence geometry [43, 46], group theory [26, 31, 45, 123], graph theory [30, 121], coding and information theory [5, 30, 36, 65, 67, 68, 99], cryptography [107, 117], as well as classification algorithms [82].

We start by introducing several notions.

Definition 1.1. An *incidence structure* is a triple $\mathcal{I} = (X, \mathcal{B}, I)$ of sets with

$$X \cap \mathcal{B} = \emptyset \ \text{ and } \ I \subseteq X \times \mathcal{B}.$$

The elements of X are called *points*, those of $\mathcal{B}$ *blocks*, and those of I *flags*.

We will usually denote points by lower-case and blocks by upper-case Latin letters. Instead of "(x, B) is a flag" it is also common to say "x and B are incident". Clearly, for a given incidence structure the role of points and blocks may be interchanged in order to obtain the *dual structure* with the given incidence relation reversed.

We will restrict ourselves in this book to *finite* incidence structures, that is, the point set X as well as the block set $\mathcal{B}$ are finite sets. Via convention, we set $v := |X|$ and $b := |\mathcal{B}|$.

For a point $x \in X$, let us define

$$(x) := \{B \in \mathcal{B} \mid (x, B) \in I\}$$

as the set of blocks incident with x, and more generally, for a subset $T \subseteq X$ of the point set

$$(T) := \{B \in \mathcal{B} \mid (x, B) \in I \text{ for each } x \in T\}.$$

Dually, for a block $B \in \mathcal{B}$, let

$$(B) := \{x \in X \mid (x, B) \in I\}$$

denote the set of points incident with B.

Incidence structures may be represented algebraically in terms of incidence matrices:

Definition 1.2. Let $\mathcal{D} = (X, \mathcal{B}, I)$ be a finite incidence structure with $|X| = v$ and $|\mathcal{B}| = b$, and let the points be labeled $\{x_1, \ldots, x_v\}$ and the blocks $\{B_1, \ldots, B_b\}$. Then, the $(v \times b)$-matrix $A = (a_{ij})$ $(1 \le i \le v, 1 \le j \le b)$ defined by

$$a_{ij} := \begin{cases} 1, & \text{if } (x_i, B_j) \in I, \\ 0, & \text{otherwise} \end{cases}$$

is called an *incidence matrix* of $\mathcal{D}$.

Clearly, A depends on the respective labeling, however, it is unique up to column and row permutation.

Incidence preserving maps which take points to points and blocks to blocks are of fundamental importance:

Definition 1.3. Let $\mathcal{I}_1 = (X_1, \mathcal{B}_1, I_1)$ and $\mathcal{I}_2 = (X_2, \mathcal{B}_2, I_2)$ be two incidence structures. A bijective map

$$\alpha : X_1 \cup \mathcal{B}_1 \longrightarrow X_2 \cup \mathcal{B}_2$$

is an *isomorphism* of $\mathcal{I}_1$ onto $\mathcal{I}_2$, if the following holds:

(i) for $x \in X_1$ and $B \in \mathcal{B}_1$, we have $x^\alpha \in X_2$ and $B^\alpha \in \mathcal{B}_2$,

(ii) for all $x \in X_1$ and all $B \in \mathcal{B}_1$, we have

$$(x, B) \in I_1 \Longleftrightarrow (x^\alpha, B^\alpha) \in I_2.$$

Two incidence structures $\mathcal{I}_1$ and $\mathcal{I}_2$ are called *isomorphic*, if there exists an isomorphism of $\mathcal{I}_1$ onto $\mathcal{I}_2$. An isomorphism of $\mathcal{I}_1$ onto itself is called an *automorphism* of $\mathcal{I}_1$. Evidently, the set of all automorphisms of an incidence structure $\mathcal{I}$ form a group under composition, the *full group of automorphisms* of $\mathcal{I}$, and will be denoted by $\mathrm{Aut}(\mathcal{I})$. Any subgroup $G \le \mathrm{Aut}(\mathcal{I})$ is called *a group of automorphisms* (or *an automorphism group*) of $\mathcal{I}$.

We will focus on those incidence structures that have certain regularity properties:

Definition 1.4. For positive integers $t \leq k \leq v$ and λ, we define a *t-design*, or more precisely a *t-(v, k, λ) design*, to be a finite incidence structure $\mathcal{D} = (X, \mathcal{B}, I)$ with the following properties:

(i) $|X| = v$,

(ii) $|(B)| = k$ for each $B \in \mathcal{B}$,

(iii) $|(T)| = \lambda$ for each t-subset $T \subseteq X$.

In other words, $\mathcal{D}$ contains v points, each block $B \in \mathcal{B}$ is incident with k points, and each t-subset of the point set X is incident with λ common blocks.

For historical reasons, a t-(v, k, λ) design with $\lambda = 1$ is called a *Steiner t-design*. Sometimes this is also known as a *Steiner system* if the parameter t is clearly given from the context. We note that in this case each block is determined by the set of points which are incident with it, and thus can be identified with a k-subset of X in a unique way.

If $t < k < v$ holds, then we speak of a *non-trivial* Steiner t-design.

The special case of a Steiner design with parameters $t = 2$ and $k = 3$ is called a *Steiner triple system of order v*. The question regarding their existence was posed in the classical "Combinatorische Aufgabe" (1853) of the nineteenth century geometer Jakob Steiner [115]:

> "Welche Zahl, N, von Elementen hat die Eigenschaft, dass sich die Elemente so zu dreien ordnen lassen, dass je zwei in einer, aber nur in einer Verbindung vorkommen?"

However, there had been earlier work on these particular designs going back to, in particular, J. Plücker, W. S. B. Woolhouse, and most notably T. P. Kirkman. For an account on the early history of designs, see [34, Chap. I.2] and [124].

A Steiner design with parameters $t = 3$ and $k = 4$ is called a *Steiner quadruple system of order v*.

If a 2-design has equally many points and blocks (that is, $v = b$), then we speak of a *square design* (as its incidence matrix is square). By tradition, square designs are often called *symmetric designs*, although here the term does not imply any symmetry of the design. A recent book on these interesting designs is due to Y. J. Ionin and M. S. Shrikhande [75].

1.2 Examples

In the following, we assume that q is a prime power and that incidence will be by natural containment. Furthermore, $GF(q)$ shall always denote the finite field with q elements.

Example 1.5 (Steiner triple system of order 7).
Let us choose as point set

$$X = \{1, 2, 3, 4, 5, 6, 7\}$$

and as block set

$$\mathcal{B} = \{\{1, 2, 4\}, \{1, 3, 7\}, \{1, 5, 6\}, \{2, 3, 5\}, \{2, 6, 7\}, \{3, 4, 6\}, \{4, 5, 7\}\}.$$

This gives a 2-$(7, 3, 1)$ design, the well-known *Fano plane*, the smallest design arising from a projective geometry, which is unique up to isomorphism. We give the usual representation of this *projective plane of order* 2 by the following diagram:

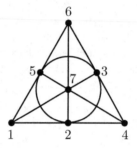

Figure 1.1: Fano plane $PG(2, 2)$

Example 1.6 (Steiner triple system of order 9).
We take as point set

$$X = \{1, 2, 3, 4, 5, 6, 7, 8, 9\}$$

and as block set

$$\mathcal{B} = \{\{1, 2, 3\}, \{4, 5, 6\}, \{7, 8, 9\}, \{1, 4, 7\}, \{2, 5, 8\}, \{3, 6, 9\},$$
$$\{1, 5, 9\}, \{2, 6, 7\}, \{3, 4, 8\}, \{1, 6, 8\}, \{2, 4, 9\}, \{3, 5, 7\}\}.$$

This gives a 2-$(9, 3, 1)$ design, the smallest non-trivial design arising from an affine geometry, which is again unique up to isomorphism. This *affine plane of order* 3 can be constructed from the array

$$\begin{array}{ccc} 1 & 2 & 3 \\ 4 & 5 & 6 \\ 7 & 8 & 9 \end{array}$$

as shown in Figure 1.2.

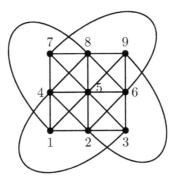

Figure 1.2: Affine plane $AG(2,3)$

More generally, we obtain:

Example 1.7 (Steiner designs from projective geometries).
We choose as point set X the set of 1-dimensional subspaces of a vector space
$V = V(d, q)$ of dimension $d \geq 3$ over $GF(q)$. As block set $\mathcal{B}$ we take the set of
2-dimensional subspaces of V. Then there are $v = (q^d - 1)/(q - 1)$ points and
each block $B \in \mathcal{B}$ is incident with $k = q + 1$ points. Since obviously any two
1-dimensional subspaces span a single 2-dimensional subspace, any two distinct
points are incident with a unique block. Thus, the projective space $PG(d - 1, q)$
is an example of a 2-$(\frac{q^d - 1}{q - 1}, q + 1, 1)$ design. For $d = 3$, the particular designs are
projective planes of order q, which are square designs.

Example 1.8 (Steiner designs from affine geometries).
We take as point set X the set of elements of a vector space $V = V(d, q)$ of
dimension $d \geq 2$ over $GF(q)$. As block set $\mathcal{B}$ we choose the set of affine lines of V
(that is, the translates of 1-dimensional subspaces). Then there are $v = q^d$ points
and each block $B \in \mathcal{B}$ is incident with $k = q$ points. As clearly any two distinct
points lie on exactly one line, they are incident with a unique block. Hence, we
obtain the affine space $AG(d, q)$ as an example of a 2-$(q^d, q, 1)$ design. When $d = 2$,
then these designs are *affine planes of order* q.

Remark 1.9. It is well-established that both affine and projective planes of order
n exist whenever n is a prime power. The conjecture that no such planes exist
with orders other than prime powers is unresolved so far. The classical result of
R. H. Bruck and H. J. Ryser [18] still remains the only general statement: If $n \equiv 1$
or 2 (mod 4) and n is not equal to the sum of two squares of integers, then n
does not occur as the order of a finite projective plane. The smallest integer that

is not a prime power and not covered by the Bruck-Ryser Theorem is 10. Using substantial computer analysis, C. W. H. Lam, L. Thiel, and S. Swiercz [88] proved the non-existence of a projective plane of order 10. The next smallest number to consider is 12, for which neither a positive nor a negative answer has been proved. Needless to mention that – apart from the existence problem – the question on the number of different isomorphism types (when existent) is fundamental. There are, for example, precisely four non-isomorphic projective planes of order 9. For a further discussion, in particular of the rich history of affine and projective planes, we refer, e.g., to [13, 43, 57, 69, 97, 108].

Specifically, we will be interested in Steiner t-designs with $t \geq 3$.

Example 1.10 (Steiner quadruple system from a cube).
We take as points the vertices of a 3-dimensional cube. As illustrated in Figure 1.3, we can choose three types of blocks:

(i) a face (six of these),

(ii) two opposite edges (six of these),

(iii) an inscribed regular tetrahedron (two of these).

This gives a 3-$(8, 4, 1)$ design, which is unique up to isomorphism.

Figure 1.3: Steiner quadruple system of order 8

We have more generally:

Example 1.11 (Steiner quadruple systems from affine geometries).
In $AG(d, q)$ any three distinct points define a plane unless they are collinear (that is, lie on the same line). If the underlying field is $GF(2)$, then the lines contain only two points and hence any three points cannot be collinear. Therefore, the points and planes of the affine space $AG(d, 2)$ form a 3-$(2^d, 4, 1)$ design.

We will see later that there are more "classical" examples of Steiner t-designs with $t \geq 3$ (such as spherical geometries), but as their construction involves some knowledge of automorphisms this will be postponed to Chapter 4. For further references concerning in particular t-designs with $t \geq 3$, we refer to [34, Chap. II.4], [70, Chap. 4], and [9, 89].

We mention that, in general, for $t = 2$ and 3, there are many infinite classes of Steiner t-designs, but for $t = 4$ and 5 only a finite number are known. Although L. Teirlinck [119] has shown that non-trivial t-designs exist for all values of t, no Steiner t-designs have been constructed for $t \geq 6$ so far.

Problem 1.12. Does there exist any non-trivial Steiner 6-design?

1.3 Basic Properties and Fisher's Inequality

For $x \in \mathbb{R}$, let $\lfloor x \rfloor$ (respectively $\lceil x \rceil$) denote the greatest positive integer which is at most (respectively the smallest positive integer which is at least) x.

All other notation remains as previously defined.

If $\mathcal{D} = (X, \mathcal{B}, I)$ is a t-(v, k, λ) design with $t \geq 2$, and $x \in X$ arbitrary, then the *derived design with respect to* x is $\mathcal{D}_x = (X_x, \mathcal{B}_x, I_x)$, where $X_x = X \setminus \{x\}$, $\mathcal{B}_x = \{B \in \mathcal{B} : (x, B) \in I\}$ and $I_x = I \mid_{X_x \times \mathcal{B}_x}$. In this case, $\mathcal{D}$ is also called an *extension* of $\mathcal{D}_x$. Obviously, $\mathcal{D}_x$ is a $(t-1)$-$(v-1, k-1, \lambda)$ design.

We will now give some helpful combinatorial tools on which we rely in the sequel.

Lemma 1.13. *Let $\mathcal{D} = (X, \mathcal{B}, I)$ be a t-(v, k, λ) design. For a positive integer $s \leq t$, let $S \subseteq X$ with $|S| = s$. Then the total number $\lambda_s := |(S)|$ of blocks incident with each element of S is given by*

$$\lambda_s = \lambda \frac{\binom{v-s}{t-s}}{\binom{k-s}{t-s}}.$$

In particular, for $t \geq 2$, a t-(v, k, λ) design is also an s-(v, k, λ_s) design.

Proof. We count in two ways the number of pairs (T, B), where $T \subseteq X$, $|T| = t$, and $B \in \mathcal{B}$, with $S \subseteq T \subseteq (B)$. First, each of the λ_s blocks B with $S \subseteq (B)$ gives $\binom{k-s}{t-s}$ such pairs. Second, there are $\binom{v-s}{t-s}$ subsets $T \subseteq X$ with $|T| = t$ and $S \subseteq T$, each giving λ pairs by Definition 1.4. $\qquad\square$

For historical reasons, it is customary to set $r := \lambda_1$ to be the total number of blocks incident with a given point (referring to the 'replication number' from statistical design of experiments, one of the origins of design theory).

The above elementary counting arguments give the following standard assertions.

Lemma 1.14. *Let $\mathcal{D} = (X, \mathcal{B}, I)$ be a t-(v, k, λ) design. Then the following holds:*

(a) $bk = vr$.

(b) $\binom{v}{t} \lambda = b \binom{k}{t}$.

(c) $r(k-1) = \lambda_2(v-1)$ *for* $t \geq 2$.

Since in Lemma 1.13 each λ_s must be an integer, we derive furthermore the subsequent necessary arithmetic conditions:

Lemma 1.15. *Let $\mathcal{D} = (X, \mathcal{B}, I)$ be a t-(v, k, λ) design. Then*

$$\lambda \binom{v-s}{t-s} \equiv 0 \left(\mathrm{mod} \ \binom{k-s}{t-s} \right)$$

for each positive integer $s \leq t$.

For non-trivial Steiner t-designs, lower bounds for v in terms of k and t can be indicated (see P. J. Cameron [24, Thm. 3A.4] and J. Tits [120, Proposition 2.2]).

Proposition 1.16. *If $\mathcal{D} = (X, \mathcal{B}, I)$ is a non-trivial Steiner t-design, then the following holds:*

(a) (Tits 1964): $v \geq (t+1)(k-t+1)$.

(b) (Cameron 1976): $v - t + 1 \geq (k - t + 2)(k - t + 1)$ *for $t > 2$. If equality holds, then $(t, k, v) = (3, 4, 8), (3, 6, 22), (3, 12, 112), (4, 7, 23)$, or $(5, 8, 24)$.*

Proof. We recall that for Steiner t-designs, a block can be identified with the set of points which are incident with it.

ad (a): Let $S \subseteq X$ with $|S| = t + 1$ such that S is not incident with any block. For each $T \subseteq S$ with $|T| = t$, there is a unique block B_T which is incident with T. Clearly, each such B_T is incident with $k - t$ points not in S, and any point not in S is incident with at most one such block B_T (since otherwise, two such blocks would be incident with more than $t - 1$ common points, contradicting the definition of Steiner t-designs). Hence, the union of all blocks B_T is incident with $(t + 1) + (t + 1)(k - t)$ points, proving Part (a).

ad (b): We will first show that for any non-trivial Steiner 2-design, we have $v - 1 \geq k(k-1)$, and equality holds if and only if any two blocks are incident with a common point: Let $B \in \mathcal{B}$, and $x \in X$ not incident with B. Due to Lemma 1.14 (c), there are $r = \frac{v-1}{k-1}$ blocks incident with x. By the definition of Steiner 2-designs, no two of these blocks are incident with a common point other than x, and none is incident with more than one point of B. Hence $\frac{v-1}{k-1} \geq k$. If equality holds, than any block incident with x has a common point with B, and hence any two blocks are incident with a common point. Clearly, the converse also holds.

Now applying the above result to the $(t - 2)$-nd derived design of $\mathcal{D}$ gives the desired inequality. If a non-trivial 2-$(v, k, 1)$ design with $v = k^2 - k + 1$ is extendable, then $(k + 1)k(k - 1)$ divides $(k^2 - k + 2)(k^2 - k + 1)(k^2 - k)$ in view of Lemma 1.14 (b), and thus $k + 1$ divides $(k^2 - k + 2)(k^2 - k + 1)$. Then, polynomial division with remainder yields that $k + 1$ divides 12 and therefore $k = 3, 5$ or 11. Considering further extensions, Lemma 1.15 shows that the designs with parameter sets $(t, k, v) = (3, 4, 8), (5, 8, 24)$, and $(3, 12, 112)$ cannot be extended, and the assertion follows.   □

We note that (a) is stronger for $k < 2(t - 1)$, while (b) is stronger for $k > 2(t - 1)$. For $k = 2(t - 1)$ both assert that $v \geq t^2 - 1$.

As we are in particular interested in the case when $3 \leq t \leq 6$, we deduce from (b) the following upper bound for the positive integer k.

Corollary 1.17. *Let* $\mathcal{D} = (X, \mathcal{B}, I)$ *be a non-trivial Steiner t-design with* $t = 3 + i$, *where* $i = 0, 1, 2, 3$. *Then the block size k can be estimated by*

$$k \leq \lfloor \sqrt{v} + \tfrac{3}{2} + i \rfloor.$$

Our next proposition is an important result by R. A. Fisher [47], generally known as "Fisher's Inequality":

Proposition 1.18 (Fisher 1940). *If* $\mathcal{D} = (X, \mathcal{B}, I)$ *is a non-trivial t-(v, k, λ) design with $t \geq 2$, then we have $b \geq v$, that is, there are at least as many blocks as points in $\mathcal{D}$.*

Proof. As a non-trivial t-design with $t \geq 2$ is also a non-trivial 2-design by Lemma 1.13, it is sufficient to prove the assertion for an arbitrary non-trivial 2-(v, k, λ) design $\mathcal{D}$. Let A be an incidence matrix of $\mathcal{D}$ labeled as in Definition 1.2. Clearly, the (i, k)-th entry

$$(AA^t)_{ik} = \sum_{j=1}^{b} (A)_{ij} (A^t)_{jk} = \sum_{j=1}^{b} a_{ij} a_{kj}$$

of the $(v \times v)$-matrix AA^t is the total number of blocks incident with both x_i and x_k, and is thus equal to r if $i = k$, and to λ if $i \neq k$. Hence

$$AA^t = (r - \lambda)I + \lambda J,$$

where I denotes the $(v \times v)$-unit matrix and J the $(v \times v)$-matrix with all entries equal to 1. Using elementary row and column operations, it follows easily that

$$\det(AA^t) = rk(r - \lambda)^{v-1}.$$

Thus AA^t is non-singular as $r = \lambda$ would imply $v = k$ by Lemma 1.13, yielding that the design is trivial. Therefore, the matrix AA^t has rank$(A) = v$. But, if $b < v$, then rank$(A) \leq b < v$, and thus rank$(AA^t) < v$, a contradiction. It follows that $b \geq v$, proving the claim. $\qquad \square$

We remark that equality holds exactly for square designs when $t = 2$. Obviously, the equality $v = b$ implies $r = k$ by Lemma 1.14 (a).

Chapter 2

Permutation Groups and Group Actions

2.1 Introduction

This chapter is on permutation groups and group actions, in particular the classification of the finite doubly transitive permutation groups will be stated. For further literature especially on finite group theory and permutation groups, we refer to [4, 32, 45, 73, 74, 87, 123].

We first give a short account of basic notions. We will restrict ourselves to finite groups, although most of the concepts also make sense for infinite groups.

Let X be a non-empty finite set. The set $\mathrm{Sym}(X)$ of all permutations of X with respect to the composition

$$x^{gh} := (x^g)^h \ \text{ for } \ x \in X \ \text{ and } \ g, h \in \mathrm{Sym}(X)$$

forms a group, called the *symmetric group* on X. If $X = \{1, \ldots, n\}$, then we write S_n for the *symmetric group of degree* n. Clearly, $\mathrm{Sym}(X) \cong S_n$ if and only if $|X| = n$.

A group G *acts* (or *operates*) on X, if to each element $g \in G$ a permutation $x \mapsto x^g$ of X is assigned such that

(i) $x^1 = x$ for all $x \in X$ (where $1 = 1_G$ denotes the identity element of G),

(ii) $(x^g)^h = x^{gh}$ for all $x \in X$ and all $g, h \in G$.

Evidently, these properties are fulfilled if and only if the map

$$\varphi : g \mapsto (x \mapsto x^g)$$

of G into $\mathrm{Sym}(X)$ is a group homomorphism.

In general, any homomorphism φ of G into $\mathrm{Sym}(X)$ is said to be an *action* (or a *permutation representation*) of G on X.

If $\ker(\varphi) = 1$ for the kernel of φ, then G acts *faithfully* on X; in this case, G is called a *permutation group* on X. If $\ker(\varphi) = G$, then G operates *trivially* on X. The *degree* of a permutation group is the size of X.

Let G and $\overline{G}$ be permutation groups acting on the sets X and $\overline{X}$, respectively. Then, G and $\overline{G}$ are called *permutation isomorphic*, if there exists a group isomorphism $\sigma : G \longrightarrow \overline{G}$ and a bijective map $\tau : X \longrightarrow \overline{X}$ with

$$(x^g)^\tau = (x^\tau)^{(g^\sigma)}$$

for all $x \in X$ and all $g \in G$. Essentially, this means that the groups are "the same" except for the labeling of the points.

Let G be a group acting on X. For $x \in X$, the subgroup

$$G_x := \{g \in G \mid x^g = x\}$$

denotes the *(point-)stabilizer* of x in G and the set

$$x^G := \{x^g \mid g \in G\}$$

is the *orbit* of x under G (or the *G-orbit* of x). The number of elements of an orbit is called the *length* of the orbit. For $B \subseteq X$, let

$$G_B := \{g \in G \mid B^g = B\}$$

be its *setwise stabilizer* and

$$G_{(B)} := \bigcap_{x \in B} G_x$$

its *pointwise stabilizer* in G. For convenience, if $B = \{x_1, \ldots, x_n\}$, we write $G_{x_1 \ldots x_n}$ in place of $G_{(B)}$, and for $x \in X$, we also denote $G_{xB} := G_x \cap G_B$.

For $g \in G$, let $\mathrm{Fix}_X(g) := \{x \in X \mid x^g = x\}$ be the set of *fixed points* and $\mathrm{Supp}_X(g) := \{x \in X \mid x^g \neq x\}$ the *support* of g in X.

A group G acting on X is said to be *transitive* on X, if G only has one orbit, that is, $x^G = X$ for all $x \in X$. Equivalently, G is transitive if for any two points $x, y \in X$ there exists an element $g \in G$ with $x^g = y$. For a positive integer $t \leq |X|$, we say that G is *t-transitive*, if for any two injective t-tuples $(x_1, x_2, \ldots, x_t)$ and $(y_1, y_2, \ldots, y_t)$ there exists an element $g \in G$ with $x_i{}^g = y_i$ for all $1 \leq i \leq t$. We say that G is *t-homogeneous*, if it is transitive on the set of all t-subsets of X. Clearly, t-transitive implies t-homogeneous.

Furthermore, we call G *semi-regular* if the identity is the only element that fixes any point of X. If additionally G is transitive, then G is said to be *regular* (or *sharply transitive*). Furthermore, an orbit is called *regular* if it has length $|G|$. The *rank* of a transitive permutation group is the number of orbits of the stabilizer of a point.

2.2 Doubly Transitive Permutation Groups

For our further purposes, we review the classification of all finite 2-transitive permutation groups, which itself relies on the classification of all finite simple groups (cf. [37, 50, 54, 55, 72, 78, 90, 100]). The list of groups is as follows:

Let G be a finite 2-transitive permutation group on a non-empty set X. Then G is either of

(A) Affine Type: G contains a regular normal subgroup T which is elementary Abelian of order $v = p^d$, where p is a prime. If a divides d, and if we identify G with a group of affine transformations

$$x \mapsto x^g + u$$

of $V = V(d, p)$, where $g \in G_0$ and $u \in V$, then particularly one of the following occurs:

(1) $G \leq A\Gamma L(1, p^d)$

(2) $G_0 \trianglerighteq SL(\frac{d}{a}, p^a)$, $d \geq 2a$

(3) $G_0 \trianglerighteq Sp(\frac{2d}{a}, p^a)$, $d \geq 2a$

(4) $G_0 \trianglerighteq G_2(2^a)'$, $d = 6a$

(5) $G_0 \cong A_6$ or A_7, $v = 2^4$

(6) $G_0 \trianglerighteq SL(2,3)$ or $SL(2,5)$, $v = p^2$, $p = 5, 7, 11, 19, 23, 29$ or 59, or $v = 3^4$

(7) G_0 contains a normal extraspecial subgroup E of order 2^5, and G_0/E is isomorphic to a subgroup of S_5, $v = 3^4$

(8) $G_0 \cong SL(2,13)$, $v = 3^6$,

or

(B) Almost Simple Type: G contains a simple normal subgroup N, and $N \leq G \leq \mathrm{Aut}(N)$. In particular, one of the following holds, where N and $v = |X|$ are given as follows:

(1) A_v, $v \geq 5$

(2) $PSL(d, q)$, $d \geq 2$, $v = \frac{q^d - 1}{q - 1}$, where $(d, q) \neq (2, 2), (2, 3)$

(3) $PSU(3, q^2)$, $v = q^3 + 1$, $q > 2$

(4) $Sz(q)$, $v = q^2 + 1$, $q = 2^{2e+1} > 2$ (Suzuki groups)

(5) $Re(q)$, $v = q^3 + 1$, $q = 3^{2e+1} > 3$ (Ree groups)

(6) $Sp(2d, 2)$, $d \geq 3$, $v = 2^{2d-1} \pm 2^{d-1}$

(7) $PSL(2, 11)$, $v = 11$

(8) $PSL(2,8)$, $v = 28$ (N is not 2-transitive)

(9) M_v, $v = 11, 12, 22, 23, 24$ (Mathieu groups)

(10) M_{11}, $v = 12$

(11) A_7, $v = 15$

(12) HS, $v = 176$ (Higman-Sims group)

(13) Co_3, $v = 276$.   (smallest Conway group)

For required basic properties of the listed groups, we refer, e.g., to [35], [73], [85, Chap. 2, 5], [118], and [122].

We also state the classification of all finite 3-homogeneous permutation groups, again relying on the classification of all finite simple groups (cf. [25, 50, 76, 90, 92]). The list of groups is as follows:

Let G be a finite 3-homogeneous permutation group on a non-empty set X. Then G is either of

(A) **Affine Type:** G contains a regular normal subgroup T which is elementary Abelian of order $v = 2^d$. If we identify G with a group of affine transformations

$$x \mapsto x^g + u$$

of $V = V(d, 2)$, where $g \in G_0$ and $u \in V$, then particularly one of the following occurs:

(1) $G \cong AGL(1,8)$, $A\Gamma L(1,8)$ or $A\Gamma L(1,32)$

(2) $G_0 \cong SL(d,2)$, $d \geq 2$

(3) $G_0 \cong A_7$, $v = 2^4$

or

(B) **Almost Simple Type:** G contains a simple normal subgroup N, and $N \leq G \leq \mathrm{Aut}(N)$. In particular, one of the following holds, where N and $v = |X|$ are given as follows:

(1) A_v, $v \geq 5$

(2) $PSL(2, q)$, $q > 3$, $v = q + 1$

(3) M_v, $v = 11, 12, 22, 23, 24$

(4) M_{11}, $v = 12$

It is elementary that if q is odd, then $PSL(2, q)$ is 3-homogeneous for $q \equiv 3 \pmod 4$, but not for $q \equiv 1 \pmod 4$, and hence not every group G of almost simple type satisfying (2) is 3-homogeneous on X.

Chapter 3

Number Theoretical Tools

3.1 Introduction

We collect in this chapter some elementary number-theoretical tools. In particular, Zsigmondy's theorem on primitive prime divisors and related issues will be useful for the investigations to come later in this book. Specific Diophantine equations are also of interest in the succeeding chapters, but will be studied in detail only when they specifically occur.

We specify some notation that will be used throughout this book. Let $\mathbb{N}$ denote the set of all positive integers (excluding zero). If m and n are integers and p a prime, then (m, n) is the greatest common divisor of m and n. Furthermore, we write $m \mid n$ if m divides n, and $p^m \parallel n$ if p^m divides n but p^{m+1} does not divide n. For $2 \leq q \in \mathbb{N}$, we mean by $z \perp q^n - 1$ that z divides $q^n - 1$ but not $q^m - 1$ for all $1 \leq m < n$. All other notation is standard.

3.2 Primitive Divisors and Zsigmondy's Theorem

We recall some basic properties of cyclotomic polynomials. For $d \in \mathbb{N}$, the d-th cyclotomic polynomial $\Phi_d(X)$ is defined as

$$\Phi_d(X) = \prod_{i=1}^{\varphi(d)} (X - \varepsilon_i),$$

where $\varepsilon_1, \ldots, \varepsilon_{\varphi(d)}$ are the primitive d-th roots of unity, and $\varphi(d)$ is Euler's totient function. We have (cf. [54, Thm. 3.3]):

Lemma 3.1 (Zsigmondy 1892). *Let $q \geq 2$ and $d \geq 1$ be integers. Then*

(a) $\Phi_d(q)$ *is an integer and* $\Phi_d(q) \mid q^d - 1$.

(b) *Let z be a prime divisor of $\Phi_d(q)$ and let $d = z^\varrho d_1$, where $z \nmid d_1$. Then d_1 is the multiplicative order of q modulo z. If $\varrho > 0$, then $z^2 \nmid \Phi_d(q)$, unless $z = d = 2$ and $q \equiv 3 \pmod 4$.*

We introduce the notion of a primitive divisor:

Definition 3.2. Let q be a prime power. An integer $z > 1$ is called a *q-primitive divisor* of $q^d - 1$, if $z \mid q^d - 1$ and $z \nmid q^a - 1$ for all $1 \le a < d$ with $a \mid d$.

A classical result on the existence of primitive divisors is due to K. Zsigmondy (see [127, p. 283], as well as [98] and [112] for simplified proofs):

Proposition 3.3 (Zsigmondy 1892). *Let q be a prime power and $d \ge 2$ an integer. Then there exists a q-primitive prime divisor of $q^d - 1$, except exactly in the following cases:*

(i) *$q = 2$ and $d = 6$.*

(ii) *$q = 2^i - 1$ ($i \ge 2$) is a Mersenne prime and $d = 2$.*

On the basis of this result a primitive prime divisor is also called a *Zsigmondy prime*. We note that Zsigmondy's Theorem was rediscovered independently by G. D. Birkhoff and H. S. Vandiver [14].

For integers $q \ge 2$ and $d \ge 1$, we define

$$\Phi_d^*(q) = \frac{1}{f^n}\Phi_d(q),$$

where $f = (d, \Phi_d(q))$ and f^n is the largest power of f dividing $\Phi_d(q)$ if $f \ne 1$, and $n = 1$ otherwise (cf. [54, p. 431]).

In modular representation theory, this definition gives a relationship between q-primitive prime divisors of $q^d - 1$ and irreducible subgroups of $GL(d, q)$ (see [54, Thm. 3.5]), which we will need later:

Proposition 3.4 (Hering 1974). *Let q be a prime power and $d \ge 1$ an integer. Then for any prime z the following conditions are equivalent:*

(i) *$z \mid \Phi_d^*(q)$.*

(ii) *$z \nmid q$ and d is the multiplicative order of q modulo z.*

(iii) *$z \perp q^d - 1$.*

(iv) *$GL(d, q)$ contains non-trivial z-groups, and every non-trivial z-group in $GL(d, q)$ is irreducible.*

(v) *$GF(q^d)$ contains an element of multiplicative order z which does not lie in any proper subfield containing $GF(q)$.*

(vi) *z is a q-primitive divisor of $q^d - 1$.*

Remark 3.5. For several applications it is useful to determine the cases in which $\Phi_d^*(q)$ has a particularly small value. For example, one obtains:

(a) If $\Phi_d^*(q) = 1$, then $q^d = p^d = 2^6$ or p^2 by Zsigmondy's Theorem.

(b) If $\Phi_d^*(q) = n + 1$, then $q^d = p^d = 2^4, 2^{10}, 2^{12}, 2^{18}, 3^4, 3^6, 5^6$ or p^2 (see [54, Thm. 3.9], where also further small values are examined).

Chapter 4

Highly Symmetric Steiner Designs

4.1 Introduction

We now look at Steiner designs which admit groups of automorphisms with sufficiently strong symmetry properties. One of the reasons for this consideration of highly symmetric designs is a general view that, while the existence of combinatorial objects is of interest, they are even more fascinating when they have a rich group of symmetries. As we will see in Section 4.2, various examples arise from finite geometries. These examples will reappear in the next chapters when dealing with classification results.

In Section 4.3, we consider properties of highly symmetric designs. Among these, flag-transitivity is certainly a particularly important and natural one. We recall that a flag of a t-(v, k, λ) design $\mathcal{D}$ is an incident point-block pair, that is, $x \in X$ and $B \in \mathcal{B}$ such that $(x, B) \in I$. In the following, we will call a group $G \le \mathrm{Aut}(\mathcal{D})$ of automorphisms of $\mathcal{D}$ *flag-transitive* (respectively *block-transitive, point t-transitive, point t-homogeneous*) if G acts transitively on the flags (respectively transitively on the blocks, t-transitively on the points, t-homogeneously on the points) of $\mathcal{D}$. For short, $\mathcal{D}$ is said to be, e.g., flag-transitive if $\mathcal{D}$ admits a flag-transitive group of automorphisms.

We introduce the concept of tactical decompositions with its implication on the orbit structures of groups of automorphisms of incidence structures. This consideration will lead to a well-known result of R. E. Block. As an application, we obtain that for any non-trivial t-design $\mathcal{D}$ with $t \ge 3$ the flag-transitivity of $G \le \mathrm{Aut}(\mathcal{D})$ always implies its point 2-transitivity (Proposition 4.13). This is the starting point for our classification of all flag-transitive Steiner t-designs with $t \ge 3$ in the following chapters.

Elementary, but often useful in the sequel, is the subsequent fact.

Lemma 4.1. *Let $\mathcal{D} = (X, \mathcal{B}, I)$ be a Steiner t-design. If $G \leq \mathrm{Aut}(\mathcal{D})$ acts flag-transitively on $\mathcal{D}$, then, for any $x \in X$, the division property*

$$r \mid |G_x|$$

holds, that is, the total number of blocks incident with a given point divides the order of the point stabilizer.

Proof. Let $x \in X$ arbitrary. As $G \leq \mathrm{Aut}(\mathcal{D})$ acts flag-transitively on $\mathcal{D}$, the stabilizer G_x acts transitively on the r blocks incident with x, and the claim follows by the orbit-stabilizer property.   □

4.2   Examples

We will see that various examples of highly symmetric Steiner designs arise from finite geometries. The examples will reappear in the next chapters when dealing with classification results.

     We assume in the following that q is a prime power, and incidence will be by natural inclusion unless otherwise stated.

Example 4.2 (Steiner designs from projective geometries).
The 2-$(\frac{q^d-1}{q-1}, q+1, 1)$ design whose points and blocks are the points and lines of the projective space $PG(d-1, q)$ (cf. Example 1.7) has as full group of automorphisms the projective semi-linear group $P\Gamma L(d, q)$.

Example 4.3 (Steiner designs from affine geometries).
The 2-$(q^d, q, 1)$ design whose points and blocks are the points and lines of the affine space $AG(d, q)$ (cf. Example 1.8) has as full group of automorphisms the affine semi-linear group $A\Gamma L(d, q)$, while the full group of automorphisms of the 3-$(2^d, 4, 1)$ design whose points and blocks are the points and planes of $AG(d, 2)$ (cf. Example 1.11) is the point 3-transitive group $AGL(d, 2)$.

Example 4.4 (Steiner designs from spherical geometries).
Let $d \geq 2$. As point set X we choose the elements of the projective line $GF(q^d) \cup \{\infty\}$ over $GF(q^d)$ (where ∞ denotes a symbol with $\infty \notin GF(q^d)$). The linear fractional group

$$PGL(2, q^d) = \{x \mapsto \frac{ax + b}{cx + d} : a, b, c, d \in GF(q^d), ad - bc \neq 0\}$$

acts on $GF(q^d) \cup \{\infty\}$ in the natural manner (with the usual conventions for ∞). As block set $\mathcal{B}$ we take the images of $GF(q) \cup \{\infty\}$ under $PGL(2, q^d)$. This gives a 3-$(q^d + 1, q + 1, 1)$ design with $PGL(2, q^d)$ as a point 3-transitive group of automorphisms. These spherical designs were first described by E. Witt [126].

For $d = 2$, these designs are often called *Möbius planes* (or *inversive planes*) *of order q*. Apart from the classical example for each prime power q, there is for $q = 2^{2e+1}$, $e \geq 1$, another Möbius plane with the Suzuki group $Sz(q)$ as a simple point 2-transitive group of automorphisms (cf. [93, Thm. 9.1 and 9.3].)

Since $PGL(2, q^2)$ acts point 3-transitively on $GF(q^2) \cup \{\infty\}$, clearly $PGL(2, q^2)$ acts also flag-transitively on the classical Möbius plane. However, $Sz(q)$ has order $(q^2 + 1)q^2(q - 1)$ and hence cannot act flag-transitively on its particular Möbius plane in view of Lemma 1.14 (c) and Lemma 4.1. Hence, this gives an example of a Steiner 3-design which is point 2-transitive but not flag-transitive.

Example 4.5 (Netto triple systems).
Let $q \equiv 7 \pmod{12}$ and ε a primitive sixth root of unity in $GF(q)$, that is, $\varepsilon^2 - \varepsilon + 1 = 0$ holds in $GF(q)$. Let $A\Gamma^2 L(1, q)$ denote the group of all permutations of $GF(q)$ of the form

$$x \mapsto a^2 x^\alpha + b,$$

where $a, b \in GF(q)$ with $a \neq 0$, and α a field automorphism. As point set X we choose the elements of $GF(q)$ and as block set $\mathcal{B}$ the images of $\{0, 1, \varepsilon\}$ under $A\Gamma^2 L(1, q)$. Thus, we obtain a 2-$(q, 3, 1)$ design $N(q)$, which is usually called a *Netto triple system* although it seems that they are not due to E. Netto (cf. [41, Sect. 3]). There are two different primitive sixth roots of unity in $GF(q)$, but P. C. Clapham [33, Prop. 3.3] showed that the respective Netto triple systems are isomorphic.

For $q = 7$ the Netto system $N(7)$ is obviously isomorphic to the Fano plane $PG(2, 2)$ and hence has a point 2-transitive full group of automorphisms. For $q > 7$, we have

$$\text{Aut}(N(q)) \cong A\Gamma^2 L(1, q),$$

and $A\Gamma^2 L(1, q)$ is 2-homogeneous but not 2-transitive on the points of $N(q)$ (see, e.g., [41] and [111]). Therefore, there is only for $q = 7$ a coincidence between Netto triple systems and projective spaces.

Example 4.6 (Mathieu-Witt designs).
The unique 2-$(9, 3, 1)$ design whose points and blocks are the points and lines of the affine plane $AG(2, 3)$ can be extended precisely three times to the following designs which are also unique up to isomorphism: the 3-$(10, 4, 1)$ design which is the Möbius plane of order 3 with $P\Gamma L(2, 9)$ as full group of automorphisms, and the two *Mathieu-Witt designs* 4-$(11, 5, 1)$ and 5-$(12, 6, 1)$ with the sporadic Mathieu groups M_{11} and M_{12} as point 4-transitive and point 5-transitive full groups of automorphisms, respectively.

To construct the "large" Mathieu-Witt designs one starts with the 2-$(21, 5, 1)$ design whose points and blocks are the points and lines of the projective plane $PG(2, 4)$. This can be extended also exactly three times to the following unique designs: the *Mathieu-Witt design* 3-$(22, 6, 1)$ with Aut(M_{22}) as point 3-transitive full group of automorphisms as well as the *Mathieu-Witt designs* 4-$(23, 7, 1)$ and 5-$(24, 8, 1)$ with M_{23} and M_{24} as point 4-transitive and point 5-transitive full groups of automorphisms, respectively.

The five Mathieu groups were the first sporadic simple groups and were discovered by E. Mathieu [101, 102] over one hundred years ago. They are the only finite 4- and 5-transitive permutation groups apart from the symmetric or alternating groups. The Steiner designs associated with the Mathieu groups were first constructed by both R. D. Carmichael [31] and E. Witt [125], and their uniqueness established up to isomorphism by Witt [126]. From the succeeding various alternative constructions, we want to refer to those of H. Lüneburg [96] and M. Aschbacher [3, Chap. 6].

Example 4.7 (Unitary designs).
A *unitary design* (or *unital*) of order n ($n \in \mathbb{N}$) is a 2-$(n^3 + 1, n + 1, 1)$ design. For $n = q$, the classical example is the *Hermitian unitary design* $U_H(q)$: Let $V = V(3, q^2)$ be a 3-dimensional vector space over $GF(q^2)$ with a non-degenerate Hermitian form. We choose as point set X the $q^3 + 1$ totally isotropic 1-dimensional subspaces of V and as block set $\mathcal{B}$ the sets of $q + 1$ points lying in a non-degenerate 2-dimensional subspace of V. The full group of automorphisms of $U_H(q)$ is the point 2-transitive group $P\Gamma L(3, q)$.

For $q = 3^{2e+1}$, $e \geq 0$, there arises a class of examples from the Ree groups: We choose as point set X the Sylow 3-subgroups and as block set $\mathcal{B}$ the set of involutions of the Ree group $Re(q)$, where an involution and a Sylow 3-subgroup are said to be incident if the involution normalizes the Sylow 3-subgroup. The *Ree unitary design* $U_R(q)$ has $Re(q)$ as a simple point 2-transitive group of automorphisms. For further details, we refer to H. Lüneburg [95].

R. Mathon [103] has constructed a class of cyclic Steiner 2-designs including a unitary design of order 6. This is the first example of a unitary design of order not a prime power.

Example 4.8 (Witt-Bose-Shrikhande designs).
Let $q = 2^e$, $e \geq 3$. The projective plane $PG(2, q)$ has an *oval* (sometimes also called *hyperoval*), that is, a set C of $q + 2$ points no three of which are collinear (cf. [57, Chap. 8]). We consider the incidence structure consisting of the points not on C and the lines not intersecting C. Taking its dual incidence structure gives a 2-$(\frac{q(q-1)}{2}, \frac{q}{2}, 1)$ design $W(q)$ with $P\Gamma L(d, q)$ as flag-transitive full group of automorphisms.

The first descriptions of these designs goes back to E. Witt [126] as well as to R. C. Bose and S. S. Shrikhande [17] in a different geometric manner, which we have illustrated above. An alternative construction from the group $PSL(2, q)$ is due to W. M. Kantor [77]: The points are the subgroups of $PSL(2, q)$ isomorphic to the dihedral group of order $2(q + 1)$, the blocks are the involutions of $PSL(2, q)$, and a point is incident with a block if the subgroup contains the involution.

We note that $W(8)$ is isomorphic to $U_R(3)$.

4.3 Block's Lemma and Related Results

In this section, we deal with decompositions of incidence matrices of incidence structures.

Definition 4.9. For a given real-valued $(m \times n)$-matrix A with $m, n \in \mathbb{N}$, let $\{R_1, \ldots, R_s\}$ be a partition of the set $\{x \in \mathbb{N} \mid 1 \leq x \leq m\}$ of row indices and $\{C_1, \ldots, C_r\}$ a partition of the set $\{x \in \mathbb{N} \mid 1 \leq x \leq n\}$ of column indices. If for each $1 \leq i \leq s$ and each $1 \leq j \leq r$ the submatrix

$$A_{ij} := A \mid_{R_i \times C_j}$$

has constant row sums d_{ij} and constant columns sums e_{ij}, then the family (A_{ij}) $(1 \leq i \leq s, 1 \leq j \leq r)$ is called a *tactical decomposition* of A.

The concept of tactical decompositions was introduced by P. Dembowski [42], probably with its origin going back to E. H. Moore [104].

We state a fundamental result of R. E. Block [15, Thm. 2], [16, Thm. 2.1] often known as "Block's Lemma". For the proof, we follow [8]. A slightly more general treatment is due to J. Siemons [114].

Proposition 4.10 (Block 1965). *Let (A_{ij}) $(1 \leq i \leq s, 1 \leq j \leq r)$ be a tactical decomposition of a real-valued $(m \times n)$-matrix A. Let $D = (d_{ij})$ and $E = (e_{ij})$ be the matrices of row and column sums of the A_{ij}, respectively. Then the following inequalities hold:*

(a) $r \leq \mathrm{rank}(D) + n - \mathrm{rank}(A) \leq s + n - \mathrm{rank}(A)$,

(b) $s \leq \mathrm{rank}(E) + m - \mathrm{rank}(A) \leq r + m - \mathrm{rank}(A)$.

Proof. For symmetric reasons, it suffices to consider Part (a). As in Definition 4.9, let $\{R_1, \ldots, R_s\}$ denote the partition of the set $\{x \in \mathbb{N} \mid 1 \leq x \leq m\}$ of row indices and $\{C_1, \ldots, C_r\}$ the partition of the set $\{x \in \mathbb{N} \mid 1 \leq x \leq n\}$ of column indices associated with the tactical decomposition (A_{ij}). We define the "row summation matrix" $U = (U_{kl})$ $(1 \leq k \leq n, 1 \leq l \leq r)$ by

$$U_{kl} := \begin{cases} 1, & \text{if } k \in C_l, \\ 0, & \text{if } k \notin C_l. \end{cases}$$

Obviously, $\mathrm{rank}(U) = r$. Setting $\tilde{D} := AU$, we obtain $\mathrm{rank}(\tilde{D}) = \mathrm{rank}(D)$ as $\tilde{D}$ consists of repeated rows of D. Let α and β denote the linear maps induced by A and U, respectively. Since

$$\alpha(\mathbb{R}^n) = \alpha(\beta(\mathbb{R}^r)) + \alpha(\beta(\mathbb{R}^r)^{\perp}) \text{ and } \dim(\beta(\mathbb{R}^r)^{\perp}) = n - r,$$

it follows that $\mathrm{rank}(A) \leq \mathrm{rank}(D) + n - r$. As clearly $\mathrm{rank}(D) \leq s$, the claim is established. $\square$

Tactical decompositions may be applied to get inside the orbit structures of groups of automorphisms of incidence structures.

Corollary 4.11. *Let $\mathcal{D} = (X, \mathcal{B}, I)$ be an incidence structure with incidence matrix A of rank $|X|$ over $\mathbb{R}$ and $G \leq \text{Aut}(\mathcal{D})$ a group of automorphisms of $\mathcal{D}$. Then, the number of orbits of G on the block set $\mathcal{B}$ is at least as large as the number of orbits of G on the point set X.*

Proof. It can easily be seen that the point and the block orbits of G form a tactical decomposition of A. Thus, the result follows immediately from Proposition 4.10.
□

From the proof of Fisher's inequality (Proposition 1.18), we deduce that a non-trivial 2-design always has an incidence matrix A with $\text{rank}(A) = v$. Since a non-trivial t-design with $t \geq 2$ is also a non-trivial 2-design by Lemma 1.13, the following specified formulation of Block's Lemma is more appropriate for our considerations:

Proposition 4.12 (Block 1965). *Let $\mathcal{D} = (X, \mathcal{B}, I)$ be a non-trivial t-(v, k, λ) design with $t \geq 2$. If $G \leq \text{Aut}(\mathcal{D})$ acts block-transitively on $\mathcal{D}$, then G acts point-transitively on $\mathcal{D}$.*

For a 2-$(v, k, 1)$ design $\mathcal{D}$, it is elementary that the point 2-transitivity of $G \leq \text{Aut}(\mathcal{D})$ implies its flag-transitivity. For 2-(v, k, λ) designs, this implication remains true if $(r, \lambda) = 1$ (see, e.g., [43, Chap. 2.3, Lemma 8]). However, for t-(v, k, λ) designs with $t \geq 3$, it can be deduced from Proposition 4.12 that always the converse holds (cf. [19] or [59, Lemma 2]):

Proposition 4.13 (Buekenhout 1968, Huber 2001). *Let $\mathcal{D} = (X, \mathcal{B}, I)$ be a non-trivial t-(v, k, λ) design with $t \geq 3$. If $G \leq \text{Aut}(\mathcal{D})$ acts flag-transitively on $\mathcal{D}$, then G acts point 2-transitively on $\mathcal{D}$.*

Proof. Let $x \in X$ arbitrary. As $G \leq \text{Aut}(\mathcal{D})$ acts flag-transitively on $\mathcal{D}$, obviously G_x acts block-transitively on the derived $(t-1)$-$(v-1, k-1, \lambda)$ design $\mathcal{D}_x$. Hence, G_x also acts point-transitively on $\mathcal{D}_x$ by Proposition 4.12, and the assertion follows.
□

For $t \geq 5$, the flag-transitivity of $G \leq \text{Aut}(\mathcal{D})$ has an even stronger implication due to the following assertion by P. J. Cameron and C. E. Praeger [29, Thm. 2.1], which follows from Proposition 4.12 and a combinatorial result of D. K. Ray-Chaudhuri and R. M. Wilson [110, Thm. 1].

Proposition 4.14 (Cameron and Praeger 1993). *Let $\mathcal{D} = (X, \mathcal{B}, I)$ be a t-(v, k, λ) design with $t \geq 2$. Then, the following holds:*

(a) *If $G \leq \text{Aut}(\mathcal{D})$ acts block-transitively on $\mathcal{D}$, then G also acts point $\lfloor t/2 \rfloor$-homogeneously on $\mathcal{D}$.*

(b) *If $G \leq \text{Aut}(\mathcal{D})$ acts flag-transitively on $\mathcal{D}$, then G also acts point $\lfloor (t+1)/2 \rfloor$-homogeneously on $\mathcal{D}$.*

Remark 4.15. If $G \leq \mathrm{Aut}(\mathcal{D})$ acts flag-transitively on any Steiner t-design $\mathcal{D}$ with $t \geq 3$, then applying Lemma 1.14 (b) and Proposition 4.13 gives the equation

$$b = \frac{\binom{v}{t}}{\binom{k}{t}} = \frac{v(v-1)\,|G_{xy}|}{|G_B|},$$

where x and y are two distinct points in X and B is a block in $\mathcal{B}$, and thus

$$\binom{v-2}{t-2} = (k-1)\binom{k-2}{t-2}\frac{|G_{xy}|}{|G_{xB}|} \quad \text{if } x \in B.$$

In the subsequent classification of all flag-transitive Steiner t-designs with $3 \leq t \leq 6$ these equations play a crucial role. In some of the cases under consideration immediately strong results are obtained. However, in some cases, particular Diophantine equations arise which have to be examined in more detail.

Chapter 5

A Census of Highly Symmetric Steiner Designs

5.1 Introduction

In this chapter, we present the complete determination of all flag-transitive Steiner t-designs with $t \geq 3$. Moreover, we survey some of the most general results on highly symmetric Steiner t-designs, without attempting to be encyclopedic. For a detailed description of the respective designs with their groups of automorphisms and for further surveys concerning in particular highly symmetric Steiner 2-designs, we refer to Chapter 4 as well as to [22, Sect. 1, 2], [43, Chap. 2.3, 2.4, 4.4], [79], [81], and [125].

5.2 Multiple Point-transitive Steiner Designs

In a beautiful classical work, T. G. Ostrom and A. Wagner [106] showed that a finite projective plane admitting a doubly point-transitive collineation group must be Desarguesian. Remarkably this succeeded – in the year 1959 clearly not possible differently – without the classification of the finite simple groups. The progress was based on an ingenious analysis of the subplane structure of projective planes. Since then the characterization of geometric or combinatorial structures in terms of their groups of automorphisms has become very popular and is now commonly viewed as a natural generalization of F. Klein's [86] Erlangen program (1872).

As probably one of the first most general results on Steiner designs, all point 2-transitive Steiner 2-designs were classified by W. M. Kantor [78, Thm. 1], using the classification of the finite 2-transitive permutation groups.

Theorem 5.1 (Kantor 1985). *Let* $\mathcal{D} = (X, \mathcal{B}, I)$ *be a non-trivial Steiner 2-design, and let* $G \leq \mathrm{Aut}(\mathcal{D})$ *act point 2-transitively on* $\mathcal{D}$. *Then one of the following holds:*

(1) $\mathcal{D}$ *is isomorphic to the* 2-$(\frac{q^d-1}{q-1}, q+1, 1)$ *design whose points and blocks are the points and lines of the projective space* $PG(d-1, q)$, *and* $PSL(d, q) \leq G \leq P\Gamma L(d, q)$, *or* $(d-1, q) = (3, 2)$ *and* $G \cong A_7$;

(2) $\mathcal{D}$ *is isomorphic to a Hermitian unital* $U_H(q)$ *of order* q, *and* $PSU(3, q^2) \leq G \leq P\Gamma U(3, q^2)$;

(3) $\mathcal{D}$ *is isomorphic to a Ree unital* $U_R(q)$ *of order* q *with* $q = 3^{2e+1} > 3$, *and* $Re(q) \leq G \leq \mathrm{Aut}(Re(q))$;

(4) $\mathcal{D}$ *is isomorphic to the* 2-$(q^d, q, 1)$ *design whose points and blocks are the points and lines of the affine space* $AG(d, q)$, *and one of the following holds:*

  (i) $G \leq A\Gamma L(1, q^d)$,

  (ii) $G_0 \trianglerighteq SL(\frac{d}{a}, q^a)$, $d \geq 2a$,

  (iii) $G_0 \trianglerighteq Sp(\frac{2d}{a}, q^a)$, $d \geq 2a$,

  (iv) $G_0 \trianglerighteq G_2(q^a)'$, q *even*, $d = 6a$,

  (v) $G_0 \trianglerighteq SL(2, 3)$ *or* $SL(2, 5)$, $v = q^2$, $q = 5, 7, 9, 11, 19, 23, 29$ *or* 59,

  (vi) $G_0 \trianglerighteq SL(2, 5)$, *or* G_0 *contains a normal extraspecial subgroup* E *of order* 2^5 *and* G_0/E *is isomorphic to a subgroup of* S_5, $v = 3^4$,

  (vii) $G_0 \cong SL(2, 13)$, $v = 3^6$;

(5) $\mathcal{D}$ *is isomorphic to the affine nearfield plane* A_9 *of order* 9, *and* G_0 *as in* (4)(vi);

(6) $\mathcal{D}$ *is isomorphic to the affine Hering plane* A_{27} *of order* 27, *and* G_0 *as in* (4)(vii);

(7) $\mathcal{D}$ *is isomorphic to one of the two Hering spaces* 2-$(9^3, 9, 1)$, *and* G_0 *as in* (4)(vii).

Moreover, for point t-transitive Steiner t-designs with $t > 2$, W. M. Kantor [78, Thm. 3] showed that the classification of the finite 2-transitive permutation groups and Theorem 5.1 easily imply the following characterization:

Theorem 5.2 (Kantor 1985). *Let* $\mathcal{D} = (X, \mathcal{B}, I)$ *be a non-trivial Steiner t-design with* $t \geq 3$, *and let* $G \leq \mathrm{Aut}(\mathcal{D})$ *act point t-transitively on* $\mathcal{D}$. *Then one of the following holds:*

(1) $\mathcal{D}$ *is isomorphic to the* 3-$(2^d, 4, 1)$ *design whose points and blocks are the points and planes of the affine space* $AG(d, 2)$, *and*

  (i) $d \geq 3$, *and* $G \cong AGL(d, 2)$, *or*

(ii) $d = 4$, *and* $G_0 \cong A_7$;

(2) $\mathcal{D}$ *is isomorphic to a* 3-$(q^e + 1, q + 1, 1)$ *design whose points are the elements of the projective line* $GF(q^e) \cup \{\infty\}$ *and whose blocks are the images of* $GF(q) \cup \{\infty\}$ *under* $PGL(2, q^e)$ *(respectively* $PSL(2, q^e)$, e *odd) with a prime power* $q \geq 3$, $e \geq 2$, *and the derived design at any given point is isomorphic to the* 2-$(q^e, q, 1)$ *design whose points and blocks are the points and lines of* $AG(e, q)$, *and* $PSL(2, q^e) \leq G \leq P\Gamma L(2, q^e)$;

(3) $\mathcal{D}$ *is isomorphic to one of the following Mathieu-Witt designs:*

  (i) *the* 3-$(22, 6, 1)$ *design, and* $G \trianglerighteq M_{22}$,

  (ii) *the* 4-$(11, 5, 1)$ *design, and* $G \cong M_{11}$,

  (iii) *the* 4-$(23, 7, 1)$ *design, and* $G \cong M_{23}$,

  (iv) *the* 5-$(12, 6, 1)$ *design, and* $G \cong M_{12}$,

  (v) *the* 5-$(24, 8, 1)$ *design, and* $G \cong M_{24}$.

5.3 Flag-transitive Steiner Designs

Among the highly symmetric properties of incidence structures, flag-transitivity is certainly a particularly important and natural one. Long before the classification of the finite simple groups, a general study of flag-transitive Steiner 2-designs was introduced by D. G. Higman and J. E. McLaughlin [56] proving that a flag-transitive group $G \leq \mathrm{Aut}(\mathcal{D})$ of automorphisms of a Steiner 2-design $\mathcal{D}$ is necessarily primitive on the points of $\mathcal{D}$. They posed the problem of classifying all finite flag-transitive projective planes, and showed that such planes are Desarguesian if their orders are suitably restricted. Much later W. M. Kantor [80] determined all such planes apart from the still open case when the group of automorphisms is a Frobenius group of prime degree. His proof involves detailed knowledge of primitive permutation groups of odd degree based on the classification of the finite simple groups.

In a big common effort, F. Buekenhout, A. Delandtsheer, J. Doyen, P. B. Kleidman, M. W. Liebeck, and J. Saxl [23, 40, 84, 91, 113] essentially characterized all finite flag-transitive linear spaces, i.e., flag-transitive Steiner 2-designs. Their result, which also relies on the classification of the finite simple groups, starts with the result of Higman and McLaughlin and uses the O'Nan-Scott Theorem for finite primitive permutation groups. For the incomplete case with a 1-dimensional affine group of automorphisms, we refer to [23, Sect. 4], [81, Sect. 3], and [21].

Theorem 5.3 (Buekenhout et al. 1990). *Let* $\mathcal{D} = (X, \mathcal{B}, I)$ *be a Steiner 2-design, and let* $G \leq \mathrm{Aut}(\mathcal{D})$ *act flag-transitively on* $\mathcal{D}$. *Then one of the following occurs:*

(1) $\mathcal{D}$ *is isomorphic to the* 2-$(q^d, q, 1)$ *design whose points and blocks are the points and lines of the affine space* $AG(d, q)$, *and one of the following holds:*

(i) G is 2-transitive (hence as in Theorem 5.1 (4)),

(ii) $d = 2$, $q = 11$ or 23, and G is one of the three solvable flag-transitive groups given in [48, Table II],

(iii) $d = 2$, $q = 9, 11, 19, 29$ or 59, $G_0^{(\infty)} \cong SL(2, 5)$ (where $G_0^{(\infty)}$ denotes the last term in the derived series of G_0), and G is given in [48, Table II],

(iv) $d = 4$, $q = 3$, and $G_0 \cong SL(2, 5)$;

(2) $\mathcal{D}$ is isomorphic to a non-Desargues affine translation plane. More precisely, one of the following holds:

(i) $\mathcal{D}$ is isomorphic to a Lüneburg-Tits plane $\mathrm{Lue}(q^2)$ of order q^2 with $q = 2^{2e+1} > 2$, and $Sz(q) \leq G_0 \leq \mathrm{Aut}(Sz(q))$,

(ii) $\mathcal{D}$ is isomorphic to the affine Hering plane A_{27} of order 27, and $G_0 \cong SL(2, 13)$,

(iii) $\mathcal{D}$ is isomorphic to the affine nearfield plane A_9 of order 9, and G is one of the seven flag-transitive subgroups of $\mathrm{Aut}(A_9)$, described in [49, §5];

(3) $\mathcal{D}$ is isomorphic to one of the two Hering spaces 2-$(9^3, 9, 1)$, and $G_0 \cong SL(2, 13)$;

(4) $\mathcal{D}$ is isomorphic to the 2-$(\frac{q^d-1}{q-1}, q+1, 1)$ design whose points and blocks are the points and lines of the projective space $PG(d-1, q)$, and $PSL(d, q) \leq G \leq P\Gamma L(d, q)$, or $(d-1, q) = (3, 2)$ and $G \cong A_7$;

(5) $\mathcal{D}$ is isomorphic to a Hermitian unital $U_H(q)$ of order q, and $PSU(3, q^2) \leq G \leq P\Gamma U(3, q^2)$;

(6) $\mathcal{D}$ is isomorphic to a Ree unital $U_R(q)$ of order q with $q = 3^{2e+1} > 3$, and $Re(q) \leq G \leq \mathrm{Aut}(Re(q))$;

(7) $\mathcal{D}$ is isomorphic to a Witt-Bose-Shrikhande space $W(q)$ with $q = 2^d \geq 8$, and $PSL(2, q) \leq G \leq P\Gamma L(2, q)$;

(8) $G \leq A\Gamma L(1, q)$.

Investigating t-designs $\mathcal{D}$ for arbitrary λ, but large t, P. J. Cameron and C. E. Praeger [29, Thm. 1.1 and 2.1] showed that for $t \geq 7$ the flag-transitivity, respectively for $t \geq 8$ the block-transitivity of $G \leq \mathrm{Aut}(\mathcal{D})$ implies at least its point 4-homogeneity (cf. Proposition 4.14) and proved as a consequence of the finite simple group classification the following result:

Theorem 5.4 (Cameron and Praeger 1993). Let $\mathcal{D} = (X, \mathcal{B}, I)$ be a t-(v, k, λ) design admitting a group $G \leq \mathrm{Aut}(\mathcal{D})$ of automorphisms. If $G \leq \mathrm{Aut}(\mathcal{D})$ acts block-transitively on $\mathcal{D}$ then $t \leq 7$, while if $G \leq \mathrm{Aut}(\mathcal{D})$ acts flag-transitively on $\mathcal{D}$ then $t \leq 6$.

Specifically, the determination of all flag-transitive Steiner t-designs with $3 \leq t \leq 6$ has attracted particular interest, but even the classification of all flag-transitive Steiner 3-designs has been known as "a long-standing and still open problem" (cf. [38, p. 147] and [39, p. 273]).

The first results in this regard go back to J. Tits [120, Thm. 1 and 2] in 1964. He provided two beautiful characterizations of the large Mathieu-Witt designs 3-$(22, 6, 1)$, 4-$(23, 7, 1)$, and 5-$(24, 8, 1)$. Let us assume that $\mathcal{D} = (X, \mathcal{B}, I)$ is a Steiner t-design. Then $t + 1$ points of X are called *independent* if they are not incident with the same block. From the construction of the large Mathieu-Witt designs provided by Witt [125, 126] (cf. Chapter 4) every Steiner t-design $\mathcal{D}$ is isomorphic to one of the Mathieu-Witt designs 3-$(22, 6, 1)$, 4-$(23, 7, 1)$, and 5-$(24, 8, 1)$ satisfies the following properties:

(A) The full group $\mathrm{Aut}(\mathcal{D})$ of automorphisms of $\mathcal{D}$ acts transitively on the set of ordered subsets of $\mathcal{D}$ consisting of $t + 1$ independent points;

(B) The full group $\mathrm{Aut}(\mathcal{D})$ of automorphisms of $\mathcal{D}$ acts transitively on the set of ordered subsets of $\mathcal{D}$ consisting of $t + 2$ points in which any $t + 1$ points are independent;

(C) Two blocks of $\mathcal{D}$ which are incident with at least $t - 2$ common points are incident with $t - 1$ common points.

Relying on Witt's result, Tits showed that the large Mathieu-Witt designs are "almost" characterized by the property (A) and one of the properties (B) or (C). More precisely, he proved the following two results:

Theorem 5.5 (Tits 1964). *Let* $\mathcal{D} = (X, \mathcal{B}, I)$ *be a non-trivial Steiner t-design with* $t \geq 2$. *Then* $\mathcal{D}$ *has the properties (A) and (C) if and only if one of the following holds:*

(1) $\mathcal{D}$ *is isomorphic to the* 2-$(q^2 + q + 1, q + 1, 1)$ *design whose points and blocks are the points and lines of the projective plane* $PG(2, q)$;

(2) $\mathcal{D}$ *is isomorphic to the* 3-$(8, 4, 1)$ *design whose points and blocks are the points and planes of the affine space* $AG(3, 2)$;

(3) $\mathcal{D}$ *is isomorphic to one of the Mathieu-Witt designs* 3-$(22, 6, 1)$, 4-$(23, 7, 1)$, *and* 5-$(24, 8, 1)$.

Theorem 5.6 (Tits 1964). *Let* $\mathcal{D} = (X, \mathcal{B}, I)$ *be a non-trivial Steiner t-design with* $t \geq 2$. *Then* $\mathcal{D}$ *has the properties (A) and (B) if and only if one of the following holds:*

(1) $\mathcal{D}$ *is isomorphic to the* 2-$(q^2 + q + 1, q + 1, 1)$ *design whose points and blocks are the points and lines of the projective plane* $PG(2, q)$;

(2) $\mathcal{D}$ *is isomorphic to the* 3-$(2^d, 4, 1)$ *design whose points and blocks are the points and planes of the affine space* $AG(d, 2)$ *with* $d \geq 3$;

(3) $\mathcal{D}$ is isomorphic to one of the Mathieu-Witt designs 3-$(22, 6, 1)$, 4-$(23, 7, 1)$, and 5-$(24, 8, 1)$.

In addition, H. Lüneburg [94] dealt in 1965 with part of the above problem characterizing flag-transitive Steiner quadruple systems under the additional strong assumption that every non-identity element of the group of automorphisms fixes at most two distinct points. We present a generalization of his result in Chapter 6, omitting the additional assumption concerning the number of fixed points.

This research monograph is devoted to the complete classification of all flag-transitive Steiner t-designs for each of the remaining parameters $t = 3, 4, 5, 6$. Besides the work of Lüneburg, the obtained results generalize the above theorems of Tits (see also [20]). Again, the primary objects that are characterized are the Mathieu-Witt designs associated with the five sporadic simple Mathieu groups; thus the results are also important for a future unified geometric theory of the sporadic simple groups (cf. [20]). Our main results have been published recently [59, 60, 61, 62, 63], and are presented here in a sufficiently self-contained and unified manner. The proofs will be given in detail in Chapters 6–10. They rely on the classification of the finite 2-transitive permutation groups and involve along with group theory also incidence geometric, combinatorial and number theoretical arguments. Especially for the latter, the study of Diophantine equations, in particular Thue-Mahler and generalized Ramanujan-Nagell equations, turns out to be helpful for crucial parts of the proofs.

Including Theorem 5.4, the complete determination of all non-trivial Steiner t-designs with $t \geq 3$ admitting a flag-transitive group of automorphisms can be stated as follows:

Theorem 5.7 (Huber 2005/07). Let $\mathcal{D} = (X, \mathcal{B}, I)$ be a non-trivial Steiner t-design with $t \geq 3$. Then $G \leq \mathrm{Aut}(\mathcal{D})$ acts flag-transitively on $\mathcal{D}$ if and only if one of the following occurs:

(1) $\mathcal{D}$ is isomorphic to the 3-$(2^d, 4, 1)$ design whose points and blocks are the points and planes of the affine space $AG(d, 2)$, and one of the following holds:

(i) $d \geq 3$, and $G \cong AGL(d, 2)$,

(ii) $d = 3$, and $G \cong AGL(1, 8)$ or $A\Gamma L(1, 8)$,

(iii) $d = 4$, and $G_0 \cong A_7$,

(iv) $d = 5$, and $G \cong A\Gamma L(1, 32)$;

(2) $\mathcal{D}$ is isomorphic to a 3-$(q^e + 1, q + 1, 1)$ design whose points are the elements of the projective line $GF(q^e) \cup \{\infty\}$ and whose blocks are the images of $GF(q) \cup \{\infty\}$ under $PGL(2, q^e)$ (respectively $PSL(2, q^e)$, e odd) with a prime power $q \geq 3$, $e \geq 2$, and the derived design at any given point is isomorphic to the 2-$(q^e, q, 1)$ design whose points and blocks are the points and lines of $AG(e, q)$, and $PSL(2, q^e) \leq G \leq P\Gamma L(2, q^e)$;

(3) $\mathcal{D}$ *is isomorphic to a* 3-$(q+1,4,1)$ *design whose points are the elements of* $GF(q) \cup \{\infty\}$ *with a prime power* $q \equiv 7 \pmod{12}$ *and whose blocks are the images of* $\{0,1,\varepsilon,\infty\}$ *under* $PSL(2,q)$, *where* ε *is a primitive sixth root of unity in* $GF(q)$, *and the derived design at any given point is isomorphic to the Netto triple system* $N(q)$, *and* $PSL(2,q) \leq G \leq P\Sigma L(2,q)$;

(4) $\mathcal{D}$ *is isomorphic to one of the following Mathieu-Witt designs:*

  (i) *the* 3-$(22,6,1)$ *design, and* $G \trianglerighteq M_{22}$,

  (ii) *the* 4-$(11,5,1)$ *design, and* $G \cong M_{11}$,

  (iii) *the* 4-$(23,7,1)$ *design, and* $G \cong M_{23}$,

  (iv) *the* 5-$(12,6,1)$ *design, and* $G \cong M_{12}$,

  (v) *the* 5-$(24,8,1)$ *design, and* $G \cong PSL(2,23)$ *or* $G \cong M_{24}$.

We remark that the Steiner 3-designs in Part (1) (ii) with $G \cong AGL(1,8)$ and (iv) with $G \cong A\Gamma L(1,32)$ as well as the Steiner 5-design in Part (4) with $G \cong PSL(2,23)$ are sharply flag-transitive, and furthermore, concerning Part (4) (v), that M_{24} as the full group of automorphisms of $\mathcal{D}$ contains only one conjugacy class of subgroups isomorphic to $PSL(2,23)$.

Chapter 6

The Classification of Flag-transitive Steiner Quadruple Systems

6.1 Introduction

We recall that a Steiner quadruple system of order v is a 3-$(v, 4, 1)$ design; in the following these will be denoted by $SQS(v)$. As we have seen in Chapter 1, Example 1.11, there exists for $d \geq 2$ always the $SQS(2^d)$ consisting of the points and planes of $AG(d, 2)$.

Using several recursive constructions, H. Hanani [52] proved in 1960 the surprising result that the following condition for the existence of a $SQS(v)$ (the necessity of which is easy to see) is also sufficient:

Proposition 6.1 (Hanani 1960). *A Steiner quadruple system $SQS(v)$ of order v exists if and only if*

$$v \equiv 2 \ or \ 4 \ (\mathrm{mod}\ 6) \ (v \geq 4).$$

For $v = 8$ and $v = 10$ there exists in each case up to isomorphism exactly one $SQS(v)$, namely the one consisting of the points and planes of $AG(3, 2)$ and the Möbius plane of order 3 (Barrau [7], 1908).

For $v = 14$ we have exactly four (Mendelsohn and Hung [71], 1972) and for $v = 16$ exactly $1,054,163$ (Kaski, Östergård and Pottonen [83], 2006) distinct isomorphism types.

In this chapter, we use the classification of the finite 2-transitive permutation groups to determine all flag-transitive $SQS(v)$. As described in Section 2.2, we have to consider two types of 2-transitive permutation groups. For both, Zsigmondy's Theorem (Proposition 3.3) will be applicable. Our result generalizes a theorem of H. Lüneburg [94] in 1965 that characterizes all flag-transitive $SQS(v)$ under the additional strong assumption that every non-identity element of the automorphism group fixes at most two points. Our procedure as well as our proofs are independent of Lüneburg. We will need the result in Chapter 7 in order to classify all flag-transitive Steiner 3-designs with arbitrary block size. Finally, we consider an additional special case concerning doubly point-transitive $SQS(v)$ (Proposition 6.6).

We state the main result of this chapter in the next section and give a complete proof in the two consecutive sections. This approach will remain for the following chapters.

6.2   Main Result

The classification of all non-trivial Steiner quadruple systems admitting a flag-transitive group of automorphisms is as follows:

Theorem 6.2. Let $\mathcal{D} = (X, \mathcal{B}, I)$ be a non-trivial Steiner quadruple system $SQS(v)$ of order v. Then $G \leq \mathrm{Aut}(\mathcal{D})$ acts flag-transitively on $\mathcal{D}$ if and only if one of the following occurs:

(1) $\mathcal{D}$ is isomorphic to the $SQS(2^d)$ whose points and blocks are the points and planes of the affine space $AG(d, 2)$, and one of the following holds:

   (i) $d \geq 3$, and $G \cong AGL(d, 2)$,

   (ii) $d = 3$, and $G \cong AGL(1, 8)$ or $A\Gamma L(1, 8)$,

   (iii) $d = 4$, and $G_0 \cong A_7$,

   (iv) $d = 5$, and $G \cong A\Gamma L(1, 32)$;

(2) $\mathcal{D}$ is isomorphic to a $SQS(3^d + 1)$ whose points are the elements of the projective line $GF(3^d) \cup \{\infty\}$ and whose blocks are the images of $GF(3) \cup \{\infty\}$ under $PGL(2, 3^d)$ with $d \geq 2$ (respectively $PSL(2, 3^d)$ with $d > 1$ odd), and the derived design at any given point is isomorphic to the 2-$(3^d, 3, 1)$ design whose points and blocks are the points and lines of $AG(d, 3)$, and $PSL(2, 3^d) \leq G \leq P\Gamma L(2, 3^d)$;

(3) $\mathcal{D}$ is isomorphic to a $SQS(q+1)$ whose points are the elements of $GF(q) \cup \{\infty\}$ with a prime power $q \equiv 7 \pmod{12}$ and whose blocks are the images of $\{0, 1, \varepsilon, \infty\}$ under $PSL(2, q)$, where ε is a primitive sixth root of unity in $GF(q)$, and the derived design at any given point is isomorphic to the Netto triple system $N(q)$, and $PSL(2, q) \leq G \leq P\Sigma L(2, q)$.

Remark 6.3. The two $SQS(32)$ given in Part (1) (iv) and Part (3) are not isomorphic. This can easily be verified by considering the derived Steiner triple systems: On the one hand, we have the Netto triple system $N(31)$, on the other, we obtain in view of [43, Chap. 2.4, Thm. 34] the Steiner triple system consisting of the points and lines of $PG(4, 2)$. However, as we have seen in Chapter 4, Example 4.5, there is only a coincidence for $q = 7$ between Netto triple systems and projective spaces.

6.3 Groups of Automorphisms of Affine Type

In this section, we start with the proof of Theorem 6.2. Throughout this chapter, let $\mathcal{D} = (X, \mathcal{B}, I)$ be a non-trivial $SQS(v)$ and $G \leq \mathrm{Aut}(\mathcal{D})$ a group of automorphisms of $\mathcal{D}$. With respect to Proposition 4.13, we may restrict ourselves to the inspection of the finite 2-transitive permutation groups listed in Section 2.2, and examine successively whether $G \leq \mathrm{Aut}(\mathcal{D})$ acts flag-transitively on $\mathcal{D}$. Let us assume in this section that G is of affine type. Then G has degree $v = p^d$ and it follows from Hanani's Theorem (Proposition 6.1) that $v = 2^d$. Since we neglect trivial $SQS(v)$, we may assume that $d \geq 3$.

We first prove the following result:

Lemma 6.4. *Let $\mathcal{D} = (X, \mathcal{B}, I)$ be a $SQS(2^d)$ with $d \geq 3$, and $G \leq \mathrm{Aut}(\mathcal{D})$ contain a regular normal subgroup T which is elementary Abelian of order $v = 2^d$. If G acts flag-transitively on $\mathcal{D}$ and $|G_0| \equiv 1 \pmod 2$, then $\mathcal{D}$ is uniquely determined (up to isomorphism), and the points and blocks of $\mathcal{D}$ are the points and planes of $AG(d, 2)$.*

Proof. Since T is elementary Abelian of order 2^d, it contains subgroups of order 4. Moreover, T is the only Sylow 2-group as $|G_0| \equiv 1 \pmod 2$, and hence contains all subgroups of G of order 4. By assumption, G_B acts transitively on the points of an arbitrary $B \in \mathcal{B}$. Thus, 4 is a divisor of the order of G_B, and G_B contains at least one subgroup S of T of order 4. Then $B \in \mathcal{B}$ is an orbit of S and hence an affine plane. As $G \leq \mathrm{Aut}(\mathcal{D})$ is block-transitive, we conclude that all blocks must be affine planes. By identifying the points of $\mathcal{D}$ with the elements of T the assertion follows. $\qquad\square$

We shall now turn to the examination of those cases where $G \leq \mathrm{Aut}(\mathcal{D})$ is of affine type.

Case (1): $G \leq A\Gamma L(1, 2^d)$.

Let us assume that $G \leq \mathrm{Aut}(\mathcal{D})$ acts flag-transitively on $\mathcal{D}$. Then, Lemma 4.1 yields

$$r = \tfrac{1}{3}(2^d - 1)(2^{d-1} - 1) \,\Big|\, |G_0| \,\Big|\, |A\Gamma L(1, 2^d)_0| = |\Gamma L(1, 2^d)| = d(2^d - 1).$$

Thus $d = 3$ or 5. First, let $d = 3$. Then $|A\Gamma L(1, 8)| = |T| |\Gamma L(1, 8)| = 8 \cdot 7 \cdot 3$. Since G is 2-transitive, we have $8 \cdot 7 \mid |G|$, and hence $|G| = 8 \cdot 7$ or $8 \cdot 7 \cdot 3$.

The latter implies that $G \cong A\Gamma L(1,8)$. So, let us assume that $|G| = 8 \cdot 7$. Since $A\Gamma L(1,8)$ is solvable and G a Hall $\{2,7\}$-group, we deduce from Hall's Theorem that $G \cong AGL(1,8)$. For $d = 5$, accordingly $|G| = 32 \cdot 31$ or $32 \cdot 31 \cdot 5$, and we conclude that $G \cong A\Gamma L(1,32)$ as for $|G| = 32 \cdot 31$ Lemma 4.1 gives a contradiction.

On the contrary, we have to show that $G \cong AGL(1,8)$, $A\Gamma L(1,8)$, respectively $A\Gamma L(1,32)$, acts flag-transitively on the $SQS(8)$, respectively the $SQS(32)$, given in the Theorem. We recall that for $v = 8$ there exists (up to isomorphism) only the unique $SQS(v)$ consisting of the points and planes of $AG(3,2)$. Since $G \cong AGL(1,8)$ acts transitively on the points, it is sufficient to show that $G_0 \cong GL(1,8)$ acts transitively on the blocks incident with the point 0. As these are the 2-dimensional subspaces of the underlying vector space, we have

$$B_1 := \{0,1,t,t+1\} \neq B_1^t = \{0,t,t^2,t^2+1\} \text{ for } 1 \neq t \in GL(1,8) \cong GF(8)^*.$$

Thus $|B_1^{GL(1,8)}| \neq 1$, and as $r = 7$ the claim follows by the orbit-stabilizer property. Hence, $G \cong A\Gamma L(1,8)$ acts flag-transitively on $\mathcal{D}$ as well. For $v = 32$, we have $|G_0| = |\Gamma L(1,32)| \equiv 1 \pmod 2$ and Lemma 6.4 gives also only the $SQS(v)$ consisting of the points and planes of $AG(5,2)$. In order to see that $G_0 \cong \Gamma L(1,32)$ acts transitively on the blocks incident with 0, examine as before that $|B_1^{GL(1,32)}| \neq 1$, and therefore $|GL(1,32)_B| = 1$ for any $0 \in B \in \mathcal{B}$. Hence $|B^{\Gamma L(1,32)}| = 31$ or $31 \cdot 5$. Assuming the first implies $|\Gamma L(1,32)_B| = 5$. Let H be a cyclic group of order 5. Then $|H_B| \neq 1$. On the other hand, 5 is a 2-primitive divisor of $2^4 - 1$. Thus H has irreducible modules of degree 4 in view of Proposition 3.4. As the 5-dimensional $GF(32)H$-module is completely reducible by Maschke's Theorem, H has as irreducible modules only the trivial module and one of degree 4. But if H fixes any 2-dimensional vector subspace then, again by Maschke's Theorem, H would have as irreducible modules two 1-dimensional modules, a contradiction. Therefore, $|B^{\Gamma L(1,32)}| = 31 \cdot 5$ must hold and the claim follows as $r = 31 \cdot 5$.

Case (2): $G_0 \trianglerighteq SL(\frac{d}{a}, p^a)$, $d \geq 2a$.

For $a = 1$ we have $G \cong AGL(d,2)$. Here G is 3-transitive and the only $SQS(v)$ on which G acts is the one whose points and blocks are the points and planes of $AG(d,2)$, $d \geq 3$, in view of Theorem 5.2. Obviously, G is also flag-transitive. Since $d \geq 2a$, we can assume that a is a proper divisor of d. We will prove that here no flag-transitive $SQS(v)$ exists.

Because of Lemma 4.1, it is enough to show that r is no divisor of $|G_0|$. Clearly,

$$|SL(\tfrac{d}{a}, 2^a)| = 2^{d(\frac{d}{a}-1)/2} \prod_{i=2}^{\frac{d}{a}} (2^{ia} - 1),$$

and $[\Gamma L(\tfrac{d}{a}, 2^a) : SL(\tfrac{d}{a}, 2^a)] = |\mathrm{Aut}(GF(2^a))|\,|GF(2^a)^*| = a(2^a - 1).$

Thus, it is sufficient to show that r does not divide $a(2^a - 1)|SL(\tfrac{d}{a}, 2^a)|$. By Zsigmondy's Theorem

$$2^{d-1} - 1$$

has a 2-primitive prime divisor z with $z \perp 2^{d-1} - 1$. Obviously, $z \neq 2$. Furthermore, $z \nmid 3a$ as $z \equiv 1 \pmod{(d-1)}$ (which follows from Proposition 3.4 (ii)) and d is properly divisible by a. Therefore,

$$2^{d-1} - 1 \nmid 3a2^{d(\frac{d}{a}-1)/2} \prod_{i=1}^{\frac{d}{a}-1} (2^{ia} - 1)$$

and the claim follows.

Cases (3)–(4).

These cases can be eliminated, analogous to Case (2), using Lemma 4.1 and Zsigmondy's Theorem.

Case (5): $G_0 \cong A_6$ or A_7, $v = 2^4$.

For $G_0 \cong A_6$, Lemma 4.1 implies that G cannot act flag-transitively on any $SQS(v)$. If $G_0 \cong A_7$, then G is 3-transitive and the only $SQS(v)$ on which G acts is the one whose points and blocks are the points and planes of $AG(4,2)$ by Theorem 5.2. Clearly, we have also flag-transitivity in this case.

Cases (6)–(8). These cases cannot occur since v is no power of 2.

6.4 Groups of Automorphisms of Almost Simple Type

We consider in this section successively those cases where G is of almost simple type. The Cases $(3), (5), (8), (12)$ can easily be ruled out by hand using Lemma 4.1. Obviously, the Cases $(4), (7), (10), (11), (13)$ can be excluded in view of Hanani's Theorem.

Before we proceed, we give the following result:

Lemma 6.5. *Let $d > 3$, and let us assume that G containing $PSL(d,q)$ as simple normal subgroup operates on the projective space $PG(d-1,q)$ and that for all $g \in G$ with $|M^g \cap M| \geq 3$ we have $M^g = M$, where M is any set of points of $PG(d-1,q)$ of cardinality k with $3 \leq k \leq |\mathcal{H}|$ and $\mathcal{H}$ a hyperplane of $PG(d-1,q)$. Then, for $|M \cap \mathcal{H}| \geq 3$, we have $M \cap \mathcal{H} = M$.*

Proof. For $k = 3$ the assertion is trivial. So, we assume that $3 < k \leq |\mathcal{H}| = \frac{q^{d-1}-1}{q-1}$. The set of all translations $T(\mathcal{H})$ form an Abelian group which operates regularly on the points of $PG(d-1,q) \setminus \mathcal{H}$ by a theorem of Baer, but trivially on $\mathcal{H}$ as the central collineations fix each point of $\mathcal{H}$. Thus the claim holds if all elements of M lie in $\mathcal{H}$. If there is an element of M which does not belong to $\mathcal{H}$, then M must contain all points of $PG(d-1,q) \setminus \mathcal{H}$. Hence

$$|M| \geq \frac{q^d - 1}{q - 1} - \frac{q^{d-1} - 1}{q - 1} = \frac{q^d - q^{d-1}}{q - 1} = q^{d-1} > \frac{q^{d-1} - 1}{q - 1} = |\mathcal{H}|,$$

which contradicts our assumption $|M| \leq |\mathcal{H}|$. □

Case (1): $N = A_v$, $v \geq 5$.

Here, G is 3-transitive and by Theorem 5.2 does not act on any non-trivial $SQS(v)$.

Case (2): $N = PSL(d, q)$, $d \geq 2$, $v = \frac{q^d - 1}{q - 1}$, where $(d, q) \neq (2, 2), (2, 3)$.

We distinguish two subcases:

Case (2a): $N = PSL(2, q)$, $v = q + 1$.

Without restriction, we have here $q \geq 5$ as $PSL(2, 4) \cong PSL(2, 5)$, and $\mathrm{Aut}(N) = P\Gamma L(2, q)$. First, we suppose that G is 3-transitive. According to Theorem 5.2, we have then only the $SQS(3^d + 1)$ described in Part (2) of Theorem 6.2 (without the subcase in brackets), and $PSL(2, 3^d) \leq G \leq P\Gamma L(2, 3^d)$. Obviously, also flag-transitivity holds. As $PGL(2, q)$ is a transitive extension of $AGL(1, q)$, it is easily seen that the derived design at any given point of $GF(3^d) \cup \{\infty\}$ is isomorphic to the 2-$(3^d, 3, 1)$ design consisting of the points and lines of $AG(d, 3)$.

Now we assume that G is 3-homogeneous but not 3-transitive. As here $PSL(2, q)$ is a transitive extension of $AG^2 L(1, q)$, we can deduce from [41] that the derived design at any given point is either the affine space $AG(d, 3)$ with the lines as blocks or the Netto triple system $N(q)$. Thus, Part (2) with the subcase in brackets or Part (3) of Theorem 6.2 holds with $PSL(2, 3^d) \leq G \leq P\Sigma L(2, 3^d)$ (where, for an odd prime p, we define $P\Sigma L(2, p^a) := PSL(2, p^a) \rtimes \langle \tau_\alpha \rangle$ with $\tau_\alpha \in \mathrm{Sym}(GF(p^a) \cup \{\infty\}) \cong S_v$ of order a induced by the Frobenius automorphism $\alpha : GF(p^a) \longrightarrow GF(p^a)$, $x \mapsto x^p$). Conversely, as G is 3-homogeneous it is also block-transitive. In both cases we have $PSL(2, q)_B \cong A_4$ for any $B \in \mathcal{B}$ as $PSL(2, q)_B$ has order 12 by the orbit-stabilizer property and $PSL(2, q)_B \longrightarrow \mathrm{Sym}(B) \cong S_4$ is a faithful representation. Thus, in both cases flag-transitivity holds.

Finally, we assume that G is not 3-homogeneous. Since $PGL(2, q)$ is 3-homogeneous the unique orbit under $PGL(2, q)$ on the 3-subsets of X splits under $PSL(2, q)$ in exactly two orbits of equal length. For each orbit, the orbit-stabilizer property gives $|PSL(2, q)_M| = |PGL(2, q)_M| = 6$ for any 3-subset M, and as $PGL(2, q)$ is 3-transitive, we have $PSL(2, q)_M \cong S_3$. If $PSL(2, q)$ acts block-transitively on any $SQS(v)$, then $PSL(2, q)_B \cong A_4$ for any $B \in \mathcal{B}$ as above. But, by the definition of $SQS(v)$, this would imply that $PSL(2, q)_{\tilde{B}}$, where $\tilde{B}$ denotes the unique block incident with M, contains $PSL(2, q)_M$, a contradiction. Thus, $PSL(2, q)$ does not act flag-transitively on any $SQS(v)$. We show now that also G cannot act flag-transitively on any $SQS(v)$. Without restriction, we may choose $\mathcal{O}_1$ to be the $PSL(2, q)$-orbit containing $\{0, 1, \infty\}$. Easy calculation shows that

$$P\Sigma L(2, q)_{0,1,\infty} = \langle \tau_\alpha \rangle.$$

Thus $P\Sigma L(2, q)_{\mathcal{O}_1}$ is contained in $P\Gamma L(2, q)$, and equality holds as $P\Sigma L(2, q)$ is of index 2 in $P\Gamma L(2, q)$ and $P\Gamma L(2, q)$ is 3-transitive. Therefore, we only have to consider

$$PSL(2, q) \leq G \leq P\Sigma L(2, q).$$

Dedekind's law gives

$$G = PSL(2,q) \rtimes (G \cap \langle \tau_\alpha \rangle).$$

Since every non-identity element of $PSL(2,q)$ fixes at most two points, we obtain

$$G_{(B)} = PSL(2,q)_{(B)} \rtimes (G \cap \langle \tau_\alpha \rangle) = G \cap \langle \tau_\alpha \rangle \cong C_m,$$

for any $B \in \mathcal{B}$, where C_m denotes the cyclic group of order $m \mid d$. If we assume that G acts block-transitively on any $SQS(v)$, we can choose $B \in \mathcal{B}$ such that B contains $\{0, 1, \infty\}$. Since $G_{(B)}$ is the kernel of the representation $G_B \longrightarrow \mathrm{Sym}(B) \cong S_4$ and $PSL(2,q)_B \cong A_4$, we have therefore again by Dedekind's law

$$G_B = PSL(2,q)_B \times (G \cap \langle \tau_\alpha \rangle) \cong A_4 \times C_m.$$

However, as $PSL(2,q)_{\{0,1,\infty\}} \cong S_3$, we get analogously

$$G_{\{0,1,\infty\}} = PSL(2,q)_{\{0,1,\infty\}} \times (G \cap \langle \tau_\alpha \rangle) \cong S_3 \times C_m,$$

contradicting the definition of a $SQS(v)$.

Case (2b): $N = PSL(d,q)$, $d \geq 3$, $v = \frac{q^d - 1}{q - 1}$.

We have here $\mathrm{Aut}(N) = P\Gamma L(d,q) \rtimes \langle \iota_\beta \rangle$, where ι_β denotes the graph automorphism induced by the inverse-transpose map $\beta : GL(d,q) \longrightarrow GL(d,q)$, $x \mapsto {}^t(x^{-1})$. We will show that G as a group of automorphisms cannot act on any non-trivial $SQS(v)$. For $d = 3$ this is obvious since $v = q^2 + q + 1$ is always odd, a contradiction to Hanani's Theorem. For $d > 3$, we will verify the claim via induction over d. For this, let us assume that there is a counter-example with d minimal. Without restriction, we can choose three distinct points x, y, z from a hyperplane $\mathcal{H}$ of $PG(d - 1, q)$. Since

$$|\mathcal{H}| = \frac{q^{d-1} - 1}{q - 1} > 4$$

for $d > 3$, Lemma 6.5 implies that the unique block $B \in \mathcal{B}$ which is incident with the 3-subset $\{x, y, z\}$ is contained completely in $\mathcal{H}$. Thus, $\mathcal{H}$ induces a $SQS(\frac{q^{d-1}-1}{q-1})$ on which G containing $PSL(d - 1, q)$ as simple normal subgroup operates. Inductively, we obtain the minimal counter-example for $d = 3$. But, as seen above, G with $PSL(3,q)$ as simple normal subgroup cannot act on any non-trivial $SQS(\frac{q^3-1}{q-1})$.

Case (6): $N = Sp(2d,2)$, $d \geq 3$, $v = 2^{2d-1} \pm 2^{d-1}$.

As here $|\mathrm{Out}(N)| = 1$, we have $N = G$. Let X^+ respectively X^- denote the set of points on which G operates. We show that G contains elements which fix exactly three points and hence by definition cannot act on any $SQS(v)$.

For a prime divisor p of $|G|$, we define

$$m_p(G) := \min\{|\mathrm{Supp}_X(g)| : 1 \neq g \text{ a } p\text{-element of } G\}$$

to be the *minimal p-degree* of a transitive permutation group G on X (cf. [58]). First, we assume that d is even. By Zsigmondy's Theorem

$$2^{d-1} - 1$$

has a 2-primitive prime divisor p with $p \perp 2^{d-1} - 1$. Moreover, p divides $|G|$ since $|G| = 2^{d^2} \prod_{i=1}^{d}(2^{2i} - 1)$. In view of [58, Thm. 3.7], we have in X^+ therefore

$$m_p(G) = 2^{2d-2(d-1)-1}(2^{2(d-1)} - 1) + 2^{d-(d-1)-1}(2^{d-1} - 1) = |X^+| - 3.$$

Thus, there exists an element $g \in G$ of prime order p that fixes exactly three points in X^+.

For $d \neq 4$, Zsigmondy's Theorem yields the existence of a 2-primitive prime divisor p with $p \perp 2^{2(d-1)} - 1$ and as p divides $|G|$, we have in X^- by [58, Thm. 3.7] again

$$m_p(G) = 2^{2d-2(d-1)-1}(2^{2(d-1)} - 1) - 2^{d-(d-1)-1}(2^{d-1} + 1) = |X^-| - 3.$$

When $d = 4$, then [35, p. 123] gives $\left|\mathrm{Fix}_{(X^-)}(g)\right| = 3$ for $g \in 3D$, where $3D$ denotes a conjugacy class in [35].

Now, let d be odd. Again by Zsigmondy's Theorem and [58, Thm. 3.7], there exists a 2-primitive prime divisor p with $p \perp 2^{2(d-1)} - 1$, and $m_p(G) = |X^-| - 3$ in X^-.

If $d \neq 7$, Zsigmondy's Theorem yields the existence of a 2-primitive prime divisor p with $p \perp 2^{d-1} - 1$. We choose $\begin{pmatrix} A_0 & A_1 \\ A_2 & A_3 \end{pmatrix} \in S \in \mathrm{Syl}_p(Sp(d-1, 2))$ and define

$$h := \begin{pmatrix} A_0 & & & A_1 & & \\ & A_0 & & & A_1 & \\ & & 1 & & & 0 \\ A_2 & & & A_3 & & \\ & A_2 & & & A_3 & \\ & & 0 & & & 1 \end{pmatrix}.$$

We deduce from [58, Thm. 3.7] that $\left|\mathrm{Fix}_{(X^+)}(h)\right| = 3$ and $\left|\mathrm{Fix}_{(X^-)}(h)\right| = 1$.

For $d = 7$, we choose $A := \begin{pmatrix} 1 & 1 \\ 1 & 0 \end{pmatrix}$ and define

$$i := \mathrm{diag}\,(A, A, A, 1, (A^{-1})^t, (A^{-1})^t, (A^{-1})^t, 1).$$

Again $\left|\mathrm{Fix}_{(X^+)}(i)\right| = 3$ and $\left|\mathrm{Fix}_{(X^-)}(i)\right| = 1$, yielding the assertion.

Case (9): M_v, $v = 11, 12, 22, 23, 24$.

Here, only $v = 22$ is possible by Hanani's Theorem. But as M_{22} is 3-transitive, Theorem 5.2 gives only the 3-$(22, 6, 1)$ design on which M_{22} respectively $\operatorname{Aut}(M_{22})$ operates.

This completes the proof of Theorem 6.2. □

In closing this chapter, we want to consider the special case when a group G of automorphisms of a non-trivial $SQS(v)$ contains $PSL(2, q)$ as simple normal subgroup. If G is 3-homogenous, then we have seen that two distinct classes of Steiner quadruple systems can occur. If G is not 3-homogeneous, which is the case if and only if $q \equiv 1 \pmod 4$, then there exists no non-trivial flag-transitive Steiner quadruple system. However, there do exist Steiner quadruple systems on which G operates point 2-transitively:

Proposition 6.6. *Let* $\mathcal{D} = (X, \mathcal{B}, I)$ *be a non-trivial Steiner quadruple system* $SQS(v)$ *of order* v. *Then* $G \le \operatorname{Aut}(\mathcal{D})$ *with* $PSL(2, q) \le G \le P\Sigma L(2, q)$, q *a prime power and* G *not 3-homogenous on* X, *operates on* $\mathcal{D}$ *if and only if* $\mathcal{D}$ *is isomorphic to the* $SQS(3^{2d} + 1)$ *whose points are the elements of* $GF(3^{2d}) \cup \{\infty\}$ *and whose blocks are the disjoint union of the images of* $\{0, 1, -1, \infty\}$ *under* $PSL(2, 3^{2d})$ *and* $\{0, 1, a, \infty\}$ *under* $PSL(2, 3^{2d})$ *with* $d \in \mathbb{N}$, $a \notin GF(3^{2d})^{*^2}$, *and* $PSL(2, 3^{2d}) \le G \le P\Sigma L(2, 3^{2d})$.

Proof. Since $v = q + 1 > 4$, we may assume that G is always a doubly transitive permutation group. Therefore, it suffices here to consider only the case when $G = PSL(2, q)$. As $PGL(2, q)$ is 3-homogeneous, the unique orbit under $PGL(2, q)$ on the 3-subsets of X splits under G in exactly two orbits of equal length. By the definition of a $SQS(v)$, it follows that G has exactly two orbits on the block set $\mathcal{B}$. By the orbit-stabilizer property, these are of equal length as for any block $B \in \mathcal{B}$ in each orbit the representation $G_B \to \operatorname{Sym}(B) \cong S_4$ is faithful, and hence

$$G_B \cong S_4.$$

We remark that G_B has then four Sylow 3-subgroups. By Hanani's Theorem and the fact that $q \equiv 1 \pmod 4$, we have to distinguish the following two cases:

(i) $q = 3^{2d}$, $d \in \mathbb{N}$.

As $3 \mid q$ each Sylow 3-subgroup has exactly one fixed point. Thus, we have at most one orbit of length 4 under G_B. On the other hand, the normalizer of a Sylow 3-subgroup in the symmetric group S_3 is S_3 itself, hence S_3 fixes the respective fixed point. The stabilizer of that fixed point in S_4 has order at least 6. But, as it is 3-closed, it cannot be S_4 itself. Moreover, it cannot be the alternating group A_4 because the latter does not contain S_3. Thus, it can only have order 6. Therefore, there exists at least one orbit of length 4. Hence, we have in each of the two orbits of blocks exactly one orbit of length 4 under G_B. This gives the circle geometries, where we choose $a \notin GF(q)^{*^2}$, since in general $-1 \in GF(q)^{*^2} \iff q \equiv 1 \pmod 4$.

As $2^4 \mid (3^d - 1)(3^d + 1)(3^{2d} + 1) = 3^{4d} - 1$ for all $d \in \mathbb{N}$, we conclude that for $q = 3^{2d}$, $d \in \mathbb{N}$, always $q^2 \equiv 1 \pmod{16}$ holds. Thus, we have in G

$$\frac{(q+1)q(q-1)}{24}$$

many subgroups isomorphic to S_4 on two conjugacy classes of equal length (cf. [44, p. 285]. As we have precisely

$$\frac{b}{2} = \frac{(q+1)q(q-1)}{48}$$

circles on each orbit of blocks, we obtain no further $SQS(v)$.

(ii) $q \equiv 1 \pmod{12}$.

Let us assume that we have an orbit of length 4 under G_B. Then, the stabilizer of a point in G_B is isomorphic to S_3. In this subgroup U isomorphic to S_3, we have a normal subgroup of order 3, which has exactly two fixed points as in particular $3 \mid q - 1$. But, as these are left fixed by an involution in U, clearly U has two fixed points. However, the stabilizer on two points in $PSL(2, q)$ is cyclic, leading to a contradiction as S_3 is non-Abelian. Hence, there cannot exist any $SQS(v)$ in this case.   □

Chapter 7

The Classification of Flag-transitive Steiner 3-Designs

7.1   Introduction

In this chapter, we completely classify all flag-transitive Steiner 3-designs with arbitrary block size. Our approach makes use of the classification of the finite 2-transitive permutation groups. Again, we have to consider two types of 2-transitive permutation groups. For the groups of affine type, nice incidence geometric arguments can be applied. In addition, for the case with a 1-dimensional affine group we use Zsigmondy's Theorem and related number theoretical results. For the groups of almost simple type, especially with the Suzuki and the Ree groups as simple normal subgroups, we have to examine the known lists of subgroups in detail for possible block stabilizers. Here, besides elementary arithmetical arguments, the study of Thue-Mahler equations, in particular generalized Ramanujan-Nagell equations, turns out to be helpful.

7.2   Main Result

The classification of all non-trivial Steiner 3-designs with a flag-transitive group of automorphisms can be stated as follows:

Theorem 7.1. *Let $\mathcal{D} = (X, \mathcal{B}, I)$ be a non-trivial Steiner 3-design. Then $G \leq \mathrm{Aut}(\mathcal{D})$ acts flag-transitively on $\mathcal{D}$ if and only if one of the following occurs:*

(1) *$\mathcal{D}$ is isomorphic to the 3-$(2^d, 4, 1)$ design whose points and blocks are the points and planes of the affine space $AG(d, 2)$, and one of the following holds:*

(i) $d \geq 3$, and $G \cong AGL(d, 2)$,

(ii) $d = 3$, and $G \cong AGL(1, 8)$ or $A\Gamma L(1, 8)$,

(iii) $d = 4$, and $G_0 \cong A_7$,

(iv) $d = 5$, and $G \cong A\Gamma L(1, 32)$;

(2) $\mathcal{D}$ is isomorphic to a 3-$(q^e + 1, q + 1, 1)$ design whose points are the elements of the projective line $GF(q^e) \cup \{\infty\}$ and whose blocks are the images of $GF(q) \cup \{\infty\}$ under $PGL(2, q^e)$ (respectively $PSL(2, q^e)$, e odd) with a prime power $q \geq 3$, $e \geq 2$, and the derived design at any given point is isomorphic to the 2-$(q^e, q, 1)$ design whose points and blocks are the points and lines of $AG(e, q)$, and $PSL(2, q^e) \leq G \leq P\Gamma L(2, q^e)$;

(3) $\mathcal{D}$ is isomorphic to a 3-$(q + 1, 4, 1)$ design whose points are the elements of $GF(q) \cup \{\infty\}$ with a prime power $q \equiv 7 \pmod{12}$ and whose blocks are the images of $\{0, 1, \varepsilon, \infty\}$ under $PSL(2, q)$, where ε is a primitive sixth root of unity in $GF(q)$, and the derived design at any given point is isomorphic to the Netto triple system $N(q)$, and $PSL(2, q) \leq G \leq P\Sigma L(2, q)$;

(4) $\mathcal{D}$ is isomorphic to the Mathieu-Witt 3-$(22, 6, 1)$ design, and $G \trianglerighteq M_{22}$.

Remark 7.2. The Steiner 3-designs in Part (1) (ii) with $G \cong AGL(1, 8)$ and (iv) with $G \cong A\Gamma L(1, 32)$ are sharply flag-transitive.

7.3   Groups of Automorphisms of Affine Type

In the following, we begin with the proof of Theorem 7.1. Let $\mathcal{D} = (X, \mathcal{B}, I)$ be a non-trivial Steiner 3-design with $G \leq \mathrm{Aut}(\mathcal{D})$ acting flag-transitively on $\mathcal{D}$ throughout this chapter. Let us recall that in view of Proposition 4.13, we can restrict ourselves to the consideration of the finite 2-transitive permutation groups listed in Section 2.2. Before we consider in this section successively those cases where G is of affine type, we prove some lemmas which will be required for Case (1).

Lemma 7.3. Let $q = p^d$ with $p \neq 2$ a prime. Furthermore, let $2^m \parallel p-1$, $2^{\overline{m}} \parallel p+1$ and $2^n \parallel d$ for some integers m, $\overline{m}$ and n. Then $2^{m+n} \parallel q-1$, unless $p \equiv 3 \pmod{4}$ and $d \equiv 0 \pmod{2}$, in which case $2^{\overline{m}+n} \parallel q - 1$.

Proof. This follows from [54, Lemma 3.2] using induction over n.   □

Maintaining the same parameters, we obtain

Lemma 7.4. Let $G \leq A\Gamma L(1, q)$ be a 2-transitive permutation group, where $q = p^d$ with $p \neq 2$ a prime, and P a Sylow 2-subgroup of G. Then we have $|P \cap AGL(1, q)| \geq 2^m$. Moreover, if $p \equiv 3 \pmod{4}$ and $d \equiv 0 \pmod{2}$, then $|P \cap AGL(1, q)| \geq 2^{\overline{m}}$.

Proof. Clearly,

$$P/P \cap AGL(1,q) \cong P \cdot AGL(1,q)/AGL(1,q) \le A\Gamma L(1,q)/AGL(1,q).$$

Thus, we obtain

$$|P| \mid |P \cap AGL(1,q)| \cdot d.$$

As $q(q-1) \mid |G|$ by the 2-transitivity of G, Lemma 7.3 yields

$$2^{m+n} \mid |P| \mid |P \cap AGL(1,q)| \cdot 2^n,$$

and therefore

$$2^m \mid |P \cap AGL(1,q)|.$$

If $p \equiv 3 \pmod 4$ and $d \equiv 0 \pmod 2$, then we have $2^{\overline{m}+n} \mid q-1$, and hence $2^{\overline{m}} \mid |P \cap AGL(1,q)|$. □

Lemma 7.5. *Let $G \le A\Gamma L(1,q)$ be a 2-transitive permutation group, where $q = p^d$ with $p \ne 2$ a prime. Then G contains an involution which fixes exactly one point.*

Proof. Clearly, $AGL(1,q)_0$ is isomorphic to $GL(1,q)$, and hence cyclic. It has index q, which is odd, and contains therefore a Sylow 2-subgroup of $AGL(1,q)$. Thus, each involution in $AGL(1,q)$ has exactly one fixed point, and the claim follows by applying Lemma 7.4. □

We shall now turn to the examination of those cases where $G \le \text{Aut}(\mathcal{D})$ is of affine type.

Case (1): $G \le A\Gamma L(1,v)$, $v = p^d$.

First, we will show by contradiction that v is a power of 2. Thus, let $p \ne 2$ and let T denote the translation subgroup of G. By Lemma 7.5, we know that G contains an involution τ which has exactly one fixed point $x \in X$. Then, for distinct $x, y \in X$, the 3-subset $S = \{x, y, y^\tau\}$ is invariant under τ. But, S is incident with a unique block $B \in \mathcal{B}$ by the definition of Steiner 3-designs, hence $\tau \in G_B$. Since G is flag-transitive, G_B acts transitively on the points of B. Therefore, for each point $x \in B$, there exists an involution τ_x having only x as fixed point. Hence

$$U := \langle \tau_x^{G_B} \rangle \le \langle \tau_x^{A\Gamma L(1,v)} \rangle = \langle \tau_x \rangle \cdot T,$$

whereas for the latter we use that τ_x induces on T the inverse map $\alpha : x \mapsto x^{-1}$ because any involutory automorphism of T which has no fixed point distinct from 1 must be equal to α. Therefore, we have $\tau_x \in AGL(1,v) \trianglelefteq A\Gamma L(1,v)$. Then, by Dedekind's law,

$$U = \langle \tau_x \rangle \cdot (U \cap T).$$

As U acts transitively on the points of B and clearly $\langle \tau_x \rangle \cap (U \cap T) = 1$, it follows from the orbit-stabilizer property that $U \cap T$ acts also transitively on the points of B. Thus, B is a point-orbit under $U \cap T$ and therefore a subspace of $AG(d, p)$. Since G is block-transitive, we conclude that all blocks must be affine subspaces.

Let $\mathcal{G}$ be a line in $AG(d, p)$ with distinct points $x, y \in \mathcal{G}$. Let B and $\overline{B}$ be two distinct blocks containing $\{x, y\}$. As $p \neq 2$ and since affine subspaces contain with any two distinct points also the line connecting them, it follows that $\mathcal{G} \subseteq B \cap \overline{B}$ with $|\mathcal{G}| > 2$, a contradiction. Thus, we have shown that $v = 2^d$.

In the following, we will prove that if the block size k is a power of 2, then only $k = 4$ can occur. Therefore, we can use the classification of all flag-transitive Steiner quadruple systems (Theorem 6.2), which gives the designs described in Part (1) of Theorem 7.1 with the assertions (ii) and (iv). Since trivial Steiner 3-designs have been excluded, let $k = 2^a$, $1 < a < d$. As $d = 3$ implies $k = 4$, we may assume that $d > 3$. From Remark 4.15, it follows that

$$v - 2 \mid d(k - 1)(k - 2). \tag{7.1}$$

Combining this with Lemma 3.1 (a) gives

$$\Phi_{d-1}^*(2) \mid 2^{d-1} - 1 \mid d(2^a - 1)(2^{a-1} - 1). \tag{7.2}$$

Clearly, $a < d - 1$ (otherwise $k = 2^{d-1}$, a contradiction to Corollary 1.17.) If $\Phi_{d-1}^*(2) = 1$, then by Proposition 3.4, there exists no non-trivial 2-primitive prime divisor of $2^{d-1} - 1$, and hence $d = 7$ in view of Zsigmondy's Theorem. By using property (7.1), Lemma 1.14 (c) and Corollary 1.17, we can easily check the very small number of possibilities for k. It turns out that only $k = 4$ can occur. Thus, we may assume that there exists a prime divisor z of $\Phi_{d-1}^*(2)$. Then $z \mid d$ by Proposition 3.4. As $z \equiv 1 \pmod{(d - 1)}$, we conclude that $z = d$. If there exists a further prime divisor $\overline{z}$ of $\Phi_{d-1}^*(2)$ with $\overline{z} \neq z$, then again $\overline{z} \mid d$ and $\overline{z} = d$ by the same arguments. Thus $\overline{z} = z$, a contradiction. Hence, we have

$$\Phi_{d-1}^*(2) = z^n$$

for some $n \in \mathbb{N}$. But then, by dividing property (7.2) by z and using Proposition 3.4 again, we obtain

$$\frac{\Phi_{d-1}^*(2)}{z} \leq 1.$$

Therefore, $\Phi_{d-1}^*(2) \leq z = d$. As $\Phi_{d-1}^*(2) = 1$ has already been considered, we may suppose that $\Phi_{d-1}^*(2) = d$. Now Remark 3.5 (b) yields $d \leq 19$. The small number of cases can easily be checked by hand as above. Again, it turns out that only $k = 4$ can occur.

Let us suppose now that k is not a power of 2. We distinguish two cases according to whether or not some non-trivial translation preserves a block $B \in \mathcal{B}$. Let $T_B \neq 1$. Then B is a disjoint union of affine subspaces X_i of $AG(d, 2)$, $i \geq 1$ (namely the point-orbits X_i of T_B contained in B). As k is not a power of 2, we may assume that $i \geq 2$. Let $x_i \in X_i$. Then the translation t mapping x_1 onto x_i maps B onto some other block B_i (because $t \notin T_B$). Since $X_i \subseteq B \cap B_i$ and

$|X_i| \geq p = 2$, it follows from the definition of Steiner 3-designs that $|X_i| = 2$ for each i. Therefore, $|G_B \cap T| = |T_B| = 2$. Without restriction, we may assume that $T_B = \langle x \mapsto x + 1 \rangle$. Thus

$$G_B \leq \mathcal{C}_{A\Gamma L(1,v)}(T_B) = T \cdot \langle \alpha \rangle,$$

where $\mathcal{C}_{A\Gamma L(1,v)}(T_B)$ denotes the centralizer of T_B in $A\Gamma L(1,v)$ and α the Frobenius automorphism $GF(v) \longrightarrow GF(v)$, $x \mapsto x^2$. Hence

$$G_B/T_B \cong G_B \cdot T/T$$

is isomorphic to a subgroup of

$$\mathcal{C}_{A\Gamma L(1,v)}(T_B)/T \cong \langle \alpha \rangle.$$

Because of the transitivity of G_B on the points of B, we conclude that $k \mid |G_B| \mid 2d$. Therefore, $v - 2 < 4d^3$ by property (7.1), and the small number of possibilities for k can easily be eliminated by hand using property (7.1) and Lemma 1.14 (c).

Now, let $T_B = 1$. We first show that $G_B \leq G_y$ for some $y \notin B$. Let $G^* = G_B \cap AGL(1, v)$. Then G^* is conjugate to a subgroup of G_0 by Hall's theorem. If $G^* = 1$, then G_B is isomorphic to a subgroup of $\langle \alpha \rangle$, hence cyclic and $|G_B| \mid d$. As G_B acts transitively on the points of B, we obtain $k \mid d$, and thus $v - 2 < d^3$ by property (7.1). The very few possibilities for k can easily be ruled out by hand as before. Therefore, $G^* \neq 1$. By construction, G^* has only the point 0 as fixed point. Since $G^* \trianglelefteq G_B$, obviously G_B fixes the set of fixed points of G^*, i.e., the point 0. Hence $G_B \leq G_0$, and $0 \notin B$ by the flag-transitivity of G.

As G is 2-transitive on points, we have $|G| = v(v-1)a$ with $a \mid d$. Then Remark 4.15 yields

$$v - 2 = (k-1)(k-2)\frac{a}{|G_{xB}|} \quad \text{if } x \in B. \tag{7.3}$$

As G_B fixes some $y \notin B$, it follows that $|G_{xB}| \mid |G_{xy}| = a$.

If G_{0x} fixes three or more distinct points, then G_{0x} would fix some block $\overline{B} \in \mathcal{B}$. Thus, we have $a \mid |G_{xB}|$, and therefore $v - 2 = (k-1)(k-2)$. However, as $d > 3$, it follows from Proposition 1.16 (b) that $v - 2 > (k-1)(k-2)$, a contradiction. Hence, G_{0x} fixes only 0 and x. Then G_{0x} must contain a field automorphism of order d, and we conclude that $G = A\Gamma L(1, 2^d)$.

Let p be a prime divisor of d, say $d = ps$. Then $(G_{0x})^p$ fixes at least three distinct points, and hence we have $s \mid |G_{xB}|$. If there exists a further prime divisor $\overline{p}$ of d with $\overline{p} \neq p$, then the quotients $d/\overline{p}$ and d/p both divide the order of G_{xB} by the flag-transitivity of G. Therefore, we obtain $d \mid |G_{xB}|$, which gives the contradiction $a = d$ as above.

Thus, we have $d = p^n$ for some $n \in \mathbb{N}$, and therefore $p^{n-1} = s \mid |G_{xB}|$. Now, it follows that $|G_{xB}| = p^{n-1}$, and hence $|G_B| = kp^{n-1} \mid (v-1)p^n$. This shows that $k \mid (v-1)p$. If we set $c = (k,p)$, then $c = 1$ or p, and we obtain $\frac{k}{c} \mid v-1$. Comparing this with equation (7.3) gives

$$v - 2 = (k-1)(k-2)\frac{p^n}{p^{n-1}},$$

and hence

$$-1 \equiv 2p \left(\operatorname{mod} \frac{k}{c}\right).$$

Therefore, we have

$$\frac{k}{c} \leq 2p + 1,$$

and finally

$$2^{p^n} - 2 = v - 2 = (k-1)(k-2)p \leq (2p^2 + p - 1)(2p^2 + p - 2)p.$$

This leaves only a small number of cases to check. As $k \mid (2^{p^n} - 1)p$, and $k \geq \left\lceil \sqrt{\frac{2^{p^n}-2}{p^n}} + \frac{3}{2} \right\rceil$ by property (7.1), these can again easily be eliminated by hand using Lemma 1.14 (c), and Corollary 1.17.

Case (2): $G_0 \unrhd SL(\frac{d}{a}, p^a)$, $d \geq 2a$.

Let e_i denote the i-th standard basis vector of the vector space $V = V(\frac{d}{a}, p^a)$, and $\langle e_i \rangle$ the 1-dimensional vector subspace spanned by e_i. We will show that only the flag-transitive designs described in Part (1) of Theorem 7.1 with $d \geq 3$ and $G \cong AGL(d,2)$ can occur.

First, let $p^a \neq 2$. For $d = 2a$, let $U = U(\langle e_1 \rangle) \leq G_0$ denote the subgroup of all transvections with axis $\langle e_1 \rangle$. Then U consists of all elements of the form

$$\begin{pmatrix} 1 & 0 \\ c & 1 \end{pmatrix}, \quad c \in GF(p^a) \text{ arbitrary}.$$

Clearly, U fixes as points only the elements of $\langle e_1 \rangle$. Hence, G_0 has point-orbits of length at least p^a outside $\langle e_1 \rangle$. Now, let $x \in \langle e_1 \rangle$ be distinct from 0 and e_1. Obviously, U fixes the unique block $B \in \mathcal{B}$ which is incident with the 3-subset $\{0, e_1, x\}$. Thus, if B contains at least one point outside $\langle e_1 \rangle$, then we would obtain $k \geq p^a + 3$. But, according to Corollary 1.17, we have $k \leq p^a + 1$, a contradiction. Therefore, B is contained completely in $\langle e_1 \rangle$. Hence, as G is flag-transitive, we may conclude that each block lies in an affine line. But, by the definition of Steiner 3-designs, any three distinct non-collinear points must also be incident with a unique block, a contradiction.

For $d \geq 3a$, we consider $(\frac{d}{a} \times \frac{d}{a})$-matrices of the form

$$A_i = \begin{pmatrix} 1 & 0 & 0 & \cdots & 0 \\ x_1 & & & & \\ 0 & & B_i & & \\ \vdots & & & & \\ 0 & & & & \end{pmatrix}, \ 1 \leq i \leq \frac{d}{a} - 1, \ x_1 \in GF(p^a) \text{ arbitrary,}$$

where

$$B_1 = \begin{pmatrix} x_2 & x_3 & x_4 & \cdots & x_{\frac{d}{a}} \\ 0 & x_2^{-1} & & & \\ 0 & & 1 & & * \\ \vdots & & 0 & & \ddots \\ 0 & & & & 1 \end{pmatrix}, \ x_2 \neq 0,$$

$$B_2 = \begin{pmatrix} 0 & x_3 & x_4 & x_5 & \cdots & x_{\frac{d}{a}} \\ x_3^{-1} & 0 & & & & \\ 0 & & -1 & & * & \\ 0 & & 1 & & & \\ \vdots & & 0 & & \ddots & \\ 0 & & & & & 1 \end{pmatrix}, \ x_3 \neq 0 \text{ and}$$

$$B_i = \begin{pmatrix} 0 & 0 & \cdots & 0 & 0 & x_{i+1} & x_{i+2} & \cdots & x_{\frac{d}{a}} \\ 0 & 1 & & & & & & & \\ \vdots & & \ddots & & & & * & & \\ 0 & & & 1 & & & & & \\ 0 & & & & -1 & & & & \\ x_{i+1}^{-1} & & & & & 0 & & & \\ 0 & & & & & & 1 & & \\ \vdots & & & & 0 & & & \ddots & \\ 0 & & & & & & & & 1 \end{pmatrix}, \ x_{i+1} \neq 0, \ 3 \leq i \leq \frac{d}{a} - 1.$$

Obviously, $B_i \in SL(\frac{d}{a} - 1, p^a)$ for $1 \leq i \leq \frac{d}{a} - 1$, and hence $A_i \in SL(\frac{d}{a}, p^a)_{e_1}$ by Laplace's expansion theorem. By multiplying e_2 with the matrices A_i $(1 \leq i \leq \frac{d}{a} - 1)$, we obtain as images exactly the vectors of $V \setminus \langle e_1 \rangle$. Thus $SL(\frac{d}{a}, p^a)_{e_1}$ and hence also G_{0,e_1} acts point-transitively on $V \setminus \langle e_1 \rangle$. Again, let $x \in \langle e_1 \rangle$ be distinct from 0 and e_1. If the unique block $B \in \mathcal{B}$ which is incident with the 3-subset $\{0, e_1, x\}$ contains some point outside $\langle e_1 \rangle$, then it would already contain all points outside, thus at least $p^d - p^a + 3$ many, which obviously contradicts Corollary 1.17. Therefore, B lies completely in $\langle e_1 \rangle$, and by the same argument as above, we obtain that here $G \leq \mathrm{Aut}(\mathcal{D})$ cannot act flag-transitively on any non-trivial Steiner 3-design $\mathcal{D}$.

Now, let $p^a = 2$. To obtain non-trivial Steiner 3-designs, let $v = 2^d > 4$. For $v = 8$, necessarily $k = 4$ must hold in view of Lemma 1.14 (c). For $v > 8$, we will show that also only Steiner quadruple systems can occur. Thus, applying Theorem 6.2 yields the claim. We remark that clearly any three distinct points are non-collinear in $AG(d, 2)$ and hence define an affine plane. Let $\mathcal{E} = \langle e_1, e_2 \rangle$ denote the 2-dimensional vector subspace spanned by e_1 and e_2. We consider $(d \times d)$-matrices of the form

$$
A_i = \begin{pmatrix}
1 & 0 & 0 & \cdots & 0 \\
0 & 1 & 0 & \cdots & 0 \\
x_1 & x_2 & & & \\
0 & 0 & & B_i & \\
\vdots & \vdots & & & \\
0 & 0 & & &
\end{pmatrix}
, \quad 1 \leq i \leq d - 2; \ x_1, x_2 \in GF(2) \text{ arbitrary}
$$

with

$$
B_1 = \begin{pmatrix}
x_3 & x_4 & x_5 & \cdots & x_d \\
0 & x_3^{-1} & & & \\
0 & & 1 & & * \\
\vdots & & & 0 & \ddots \\
0 & & & & 1
\end{pmatrix}
, \quad x_3 \neq 0,
$$

$$
B_2 = \begin{pmatrix}
0 & x_4 & x_5 & x_6 & \cdots & x_d \\
x_4^{-1} & 0 & & & & \\
0 & & -1 & & * & \\
0 & & & 1 & & \\
\vdots & & & & 0 & \ddots \\
0 & & & & & 1
\end{pmatrix}
, \quad x_4 \neq 0 \text{ and}
$$

$$
B_i = \begin{pmatrix}
0 & 0 & \cdots & 0 & 0 & x_{i+2} & x_{i+3} & \cdots & x_d \\
0 & 1 & & & & & & & \\
\vdots & & \ddots & & & & & & \\
0 & & & 1 & & & * & & \\
0 & & & & -1 & & & & \\
x_{i+2}^{-1} & & & & & 0 & & & \\
0 & & & & & & 1 & & \\
\vdots & & & & 0 & & & \ddots & \\
0 & & & & & & & & 1
\end{pmatrix}
, \quad x_{i+2} \neq 0, \ 3 \leq i \leq d - 2.
$$

Analogously as above, $B_i \in SL(d-2,2)$ for $1 \leq i \leq d-2$ and $A_i \in SL(d,2)_\mathcal{E}$. By multiplying e_3 with the matrices A_i $(1 \leq i \leq d-2)$, we obtain as images exactly the vectors of $V \setminus \mathcal{E}$. Hence $SL(d,2)_\mathcal{E}$ and therefore also $G_{0,\mathcal{E}}$ acts point-transitively on $V \setminus \mathcal{E}$. If the unique block $B \in \mathcal{B}$ which is incident with the 3-subset $\{0, e_1, e_2\}$ contains some point outside $\mathcal{E}$, then it would already contain all points of $V \setminus \mathcal{E}$. But then, we would have $k \geq 2^d - 4 + 3 = 2^d - 1$, a contradiction to Corollary 1.17. Hence, B lies completely in $\mathcal{E}$, and by the flag-transitivity of G, it follows that each block must be contained in an affine plane. Therefore, we have $k \leq 4$, and in particular $k = 4$ as trivial Steiner 3-designs are excluded.

Case (3): $G_0 \trianglerighteq Sp(\frac{2d}{a}, p^a)$, $d \geq 2a$.

We will prove by contradiction that $G \leq \text{Aut}(\mathcal{D})$ cannot act flag-transitively on any non-trivial Steiner 3-design $\mathcal{D}$. First, let $p^a \neq 2$. The permutation group $PSp(\frac{2d}{a}, p^a)$ on the points of the associated projective space is a rank 3 group, and the orbits of the one-point stabilizer are well-known (e.g. [73, Chap. II, Thm. 9.15 (b)]). Thus, $G_0 \trianglerighteq Sp(\frac{2d}{a}, p^a)$ has exactly two orbits on $V \setminus \langle x \rangle$ $(0 \neq x \in V)$ of length at least

$$\frac{p^a(p^{2d-2a}-1)}{p^a-1} = \sum_{i=1}^{\frac{2d}{a}-2} p^{ia} > p^d.$$

Let $y \in \langle x \rangle$ be distinct from 0 and x. If the unique block incident with the 3-subset $\{0, x, y\}$ contains at least one point of $V \setminus \langle x \rangle$, then we would have $k > p^d + 3$. But, on the other hand, we have $k \leq p^d + 1$ by Corollary 1.17, a contradiction. Therefore, we can argue as in Case (2) to obtain the desired contradiction.

Now, let $p^a = 2$. We may assume that $v = 2^{2d} > 4$. For $d = 2$ (here $Sp(4,2) \cong S_6$ as well-known), Corollary 1.17 yields $k \leq 5$. As $k - 2 \nmid v - 2$ for $k = 5$, it is sufficient by Lemma 1.14 (c) to consider the case when $k = 4$. For $d > 2$, we will show that we can also restrict ourselves to Steiner quadruple systems. Hence, the claim follows from Theorem 6.2 again. It is easily seen that there are $2^{2d-1}(2^{2d} - 1)$ hyperbolic pairs in the non-degenerate symplectic space $V = V(2d, 2)$, and by Witt's theorem, $Sp(2d, 2)$ is transitive on these hyperbolic pairs (cf. [73, Chap. II, Thm. 9.13]). Let $\{x, y\}$ denote a hyperbolic pair, and $\mathcal{E} = \langle x, y \rangle$ the hyperbolic plane spanned by $\{x, y\}$. As $\mathcal{E}$ is non-degenerate, we have the orthogonal decomposition

$$V = \mathcal{E} \perp \mathcal{E}^\perp.$$

Clearly, $Sp(2d, 2)_{\{x,y\}}$ stabilizes $\mathcal{E}^\perp$ as a subspace, which implies that $Sp(2d, 2)_{\{x,y\}} \cong Sp(2d - 2, 2)$. As $|\text{Out}(Sp(2d, 2))| = 1$, we have therefore

$$Sp(2d - 2, 2) \cong Sp(2d, 2)_{\{x,y\}} \trianglelefteq Sp(2d, 2)_\mathcal{E} = G_{0,\mathcal{E}}.$$

Since $Sp(2d - 2, 2)$ acts transitively on the non-zero vectors of the $(2d - 2)$-dimensional symplectic subspace, it is easy to see that the smallest orbit on $V \setminus \mathcal{E}$

under $G_{0,\varepsilon}$ has length at least $2^{2d-2} - 1$. If the unique block $B \in \mathcal{B}$ which is incident with the 3-subset $\{0, x, y\}$ contains some point in $V \setminus \mathcal{E}$, then we would have $k \geq 2^{2d-2} + 2$, a contradiction to Corollary 1.17. Thus, B lies completely in $\mathcal{E}$, and with regard to the flag-transitivity of G, we conclude that each block must be contained in an affine plane. Hence, $k = 4$ as trivial Steiner 3-designs are excluded.

Case (4): $G_0 \trianglerighteq G_2(2^a)'$, $d = 6a$.

We will also show by contradiction that $G \leq \mathrm{Aut}(\mathcal{D})$ cannot act flag-transitively on any non-trivial Steiner 3-design $\mathcal{D}$. First, let $a = 1$. Then we have $v = 2^6 = 64$, and by Corollary 1.17, it follows that $k \leq 9$. On the other hand, we have $|G_2(2)'| = 2^5 \cdot 3^3 \cdot 7$ and $|\mathrm{Out}(G_2(2)')| = 2$. Thus, in view of Lemma 4.1, we obtain

$$r = \frac{63 \cdot 62}{(k-1)(k-2)} \ \Big|\ |G_0| \ \Big|\ 2^6 \cdot 3^3 \cdot 7.$$

But this implies that $k - 1$ or $k - 2$ is a multiple of 31, a contradiction.

Now, let $a > 1$. As here $G_2(2^a)$ is simple non-Abelian, it is sufficient to consider $G_0 \trianglerighteq G_2(2^a)$. The permutation group $G_2(2^a)$ is of rank 4, and for $0 \neq x \in V$, the one-point stabilizer $G_2(2^a)_x$ has exactly three orbits $\mathcal{O}_i$ ($i = 1, 2, 3$) on $V \setminus \langle x \rangle$ of length $2^{3a} - 2^a, 2^{5a} - 2^{3a}, 2^{6a} - 2^{5a}$ (see, e.g., [2] or [27, Thm. 3.1]). Thus, G_0 has exactly three orbits on $V \setminus \langle x \rangle$ of length at least $|\mathcal{O}_i|$. Let $y \in \langle x \rangle$ be distinct from 0 and x. Again, we will show that the unique block $B \in \mathcal{B}$ which is incident with the 3-subset $\{0, x, y\}$ lies completely in $\langle x \rangle$. If B contains at least one point of $V \setminus \langle x \rangle$ in $\mathcal{O}_2$ or $\mathcal{O}_3$, then we would obtain as above a contradiction to Corollary 1.17. Thus, we only have to consider the case when B contains points of $V \setminus \langle x \rangle$ which all lie in $\mathcal{O}_1$. By [2], the orbit $\mathcal{O}_1$ is exactly known, and we have

$$\mathcal{O}_1 = x\Delta \setminus \langle x \rangle,$$

where $x\Delta = \{y \in V \mid f(x, y, z) = 0 \text{ for all } z \in V\}$ with an alternating trilinear form f on V. Then B consists, apart from elements of $\langle x \rangle$, exactly of $\mathcal{O}_1$. Since $|\mathcal{O}_1| \neq 1$, we can choose $\langle \overline{x} \rangle \in x\Delta$ with $\langle \overline{x} \rangle \neq \langle x \rangle$. Let $\overline{y} \in \langle \overline{x} \rangle$ be distinct from 0 and $\overline{x}$. Then, for symmetry reasons, the 3-subset $\{0, \overline{x}, \overline{y}\}$ is also incident with the unique block B. But, on the other hand, we have $\overline{x}\Delta \neq x\Delta$ for $\langle \overline{x} \rangle \neq \langle x \rangle$, a contradiction. Thus, B is contained completely in $\langle x \rangle$, and we may argue as in the cases above.

Case (5): $G_0 \cong A_6$ or A_7, $v = 2^4$.

As $v = 2^4$, we have $k \leq 5$ by Corollary 1.17. If $k = 4$, then applying Theorem 6.2 gives the flag-transitive design described in Part (1) of Theorem 7.1 with assertion (iii). For $k = 5$, we obtain with Lemma 1.14 (c) a contradiction.

Cases (6)–(8).

For the existence of non-trivial Steiner 3-designs, we have in these cases only a very small number of possibilities for k to check, which can easily be ruled out by hand using Lemma 1.14 (b) and (c), and Corollary 1.17.

7.4 Groups of Automorphisms of Almost Simple Type

We will examine in this section successively those cases where G is of almost simple type. We note that for each of the Cases $(7), (8), (10)$–(13) we have only a small number of possibilities for k to check, which can be eliminated by hand using Lemma 1.14 (b) and (c), Corollary 1.17, and Lemma 4.1.

Case (1): $N = A_v$, $v \geq 5$.

Here, G is 3-transitive and by Theorem 5.2 does not act on any non-trivial Steiner 3-design.

Case (2): $N = PSL(d, \tilde{q})$, $d \geq 2$, $v = \frac{\tilde{q}^d - 1}{\tilde{q} - 1}$, where $(d, \tilde{q}) \neq (2, 2), (2, 3)$.

We distinguish two subcases:

Case (2a): $N = PSL(2, \tilde{q})$, $v = \tilde{q} + 1$.

Let $\tilde{q} = q^e$, $e \geq 1$. Without restriction, we have here $q^e \geq 5$ as $PSL(2, 4) \cong PSL(2, 5)$, and $\mathrm{Aut}(N) = P\Gamma L(2, q^e)$. First, we suppose that G is 3-transitive. In view of Theorem 5.2, we have then only the 3-$(q^e + 1, q + 1, 1)$ design described in Part (2) of Theorem 7.1 (without the subcase in brackets) with $PSL(2, q^e) \leq G \leq P\Gamma L(2, q^e)$, $q \geq 3$, $e \geq 2$. Conversely, flag-transitivity holds as the 3-transitivity of G implies that G_x acts block-transitively on the derived Steiner 2-design $\mathcal{D}_x$ for any $x \in X$. Since $PGL(2, q^e)$ is a transitive extension of $AGL(1, q^e)$, it is easily seen that the derived design at any given point of $GF(q^e) \cup \{\infty\}$ is isomorphic to the 2-$(q^e, q, 1)$ design consisting of the points and lines of $AG(e, q)$.

Now, we suppose that G is 3-homogeneous but not 3-transitive. Since here $PSL(2, q^e)$ is a transitive extension of $AG^2L(1, q^e)$, we can deduce from [41] that the derived design at any given point is either $AG(e, q)$ with the lines as blocks or the Netto triple system $N(q^e)$. Thus, Part (2) of Theorem 7.1 holds with the subcase in brackets or Part (3) with $PSL(2, q^e) \leq G \leq P\Sigma L(2, q^e)$. Conversely, in view of its 3-homogeneity, G is also block-transitive. By the orbit-stabilizer property, we obtain $|PSL(2, q^e)_B| = |PSL(2, q)|$ and in view of [44, Chap. 12, p. 286] actually

$$PSL(2, q^e)_B \cong PSL(2, q)$$

for any $B \in \mathcal{B}$. Since $PSL(2, q)$ acts 2-transitively on $k = q + 1$ points, it follows that in both cases flag-transitivity holds.

Finally, we assume that G is not 3-homogeneous. As $PGL(2, q^e)$ is 3-homogeneous, the unique orbit under $PGL(2, q^e)$ on the 3-subsets of X splits under $PSL(2, q^e)$ in exactly two orbits of equal length. Thus, G has here exactly two orbits of equal length on the 3-subsets of X, and by the definition of Steiner 3-designs, it follows that G has exactly two orbits (possibly of different length) on the blocks. Hence, $G \leq \mathrm{Aut}(\mathcal{D})$ cannot act block-transitively and therefore not flag-transitively on any non-trivial Steiner 3-design $\mathcal{D}$.

Case (2b): $N = PSL(d, \tilde{q})$, $d \geq 3$, $v = \frac{\tilde{q}^d - 1}{\tilde{q} - 1}$.

We have here $\mathrm{Aut}(N) = P\Gamma L(d, \tilde{q}) \rtimes \langle \iota_\beta \rangle$, where ι_β denotes the graph automorphism induced by the inverse-transpose map $\beta : GL(d, \tilde{q}) \longrightarrow GL(d, \tilde{q})$, $x \mapsto {}^t(x^{-1})$. We will prove by contradiction that $G \leq \mathrm{Aut}(\mathcal{D})$ cannot act on any non-trivial Steiner 3-design $\mathcal{D}$.

Let us first assume that $d = 3$. By the definition of Steiner 3-designs, we may choose in the underlying projective plane $PG(2, \tilde{q})$ three distinct non-collinear points $x, y, z \in X$ which are incident with a unique block $B \in \mathcal{B}$. We consider two subcases:

(i) B contains at least one further point of the triangle through x, y, z.

(ii) B does not contain any further point of the triangle.

ad (i): Let $\mathcal{G}$ denote a line of $PG(2, \tilde{q})$. We know that the translation group $T(\mathcal{G})$ operates regularly on the points of $PG(2, \tilde{q}) \setminus \mathcal{G}$ and trivially on $\mathcal{G}$. Thus, $T(\mathcal{G})$ fixes a block $B \in \mathcal{B}$ if three or more distinct points of B lie on $\mathcal{G}$. Therefore, the block mentioned in (i) must contain all points of $PG(2, \tilde{q}) \setminus \mathcal{G}$, thus at least $\tilde{q}^2 + 3$ many. But, these are obviously more than half of the points of $PG(2, \tilde{q})$, a contradiction to $k \leq \lfloor \frac{v}{4} + 2 \rfloor$ by Proposition 1.16 (a).

ad (ii): The pointwise stabilizer of three distinct points in $SL(3, \tilde{q})$ consists precisely of the diagonal matrices, and hence has order $(\tilde{q} - 1)^2$ (see, e.g., [73, Chap. II, Thm. 7.2 (b)]). To this corresponds in $PSL(3, \tilde{q})$ a subgroup U of order

$$\tfrac{1}{n}(\tilde{q} - 1)^2 \quad \text{with } n = (3, \tilde{q} - 1).$$

As U acts semi-regularly outside the triangle, we obtain n point-orbits of equal length $\frac{1}{n}(\tilde{q} - 1)^2$, since if U fixes some further point outside the triangle, then U would fix some non-degenerate quadrangle, and so would be the identity element, a contradiction. Thus, we get

$$k \geq 3 + \tfrac{1}{n}(\tilde{q} - 1)^2.$$

On the other hand, we know that the block mentioned in (ii) is an arc, and therefore contains at most $\tilde{q} + 1$ points for $\tilde{q}$ odd or $\tilde{q} + 2$ points for $\tilde{q}$ even (see, e.g., [43, Chap. 3.2, Thm. 24]). Only for $\tilde{q} = 2$ and 4 are both conditions fulfilled. But, with regard to Lemma 1.14 (c), there exist no non-trivial 3-$(7, k, 1)$ designs and 3-$(21, k, 1)$ designs. Therefore, for $d = 3$ we have shown that G cannot act on any non-trivial 3-$(\tilde{q}^2 + \tilde{q} + 1, k, 1)$ design.

Now, we consider the case when $d > 3$. Via induction over d, we will verify that $G \leq \mathrm{Aut}(\mathcal{D})$ cannot act on any non-trivial Steiner 3-design $\mathcal{D}$. For this, let us assume that there is a counterexample with d minimal. Without restriction, we can choose three distinct points x, y, z from a hyperplane $\mathcal{H}$ of $PG(d - 1, \tilde{q})$. The translation group $T(\mathcal{H})$ acts regularly on the points of $PG(d - 1, \tilde{q}) \setminus \mathcal{H}$ and trivially on $\mathcal{H}$. Therefore, if the unique block $B \in \mathcal{B}$ which is incident with the

3-subset $\{x, y, z\}$ contains at least one point outside $\mathcal{H}$, then it would already contain all points of $PG(d-1, \tilde{q}) \setminus \mathcal{H}$, thus at least $\tilde{q}^{d-1} + 3$ many. However, as

$$v = \frac{\tilde{q}^d - 1}{\tilde{q} - 1} < 2\tilde{q}^{d-1} \Longleftrightarrow \tilde{q}^d - 1 < 2(\tilde{q}^d - \tilde{q}^{d-1}) \Longleftrightarrow 2\tilde{q}^{d-1} - 1 < \tilde{q}^d,$$

these are more than half of the points of $PG(d-1, \tilde{q})$, again a contradiction. Hence B is contained completely in $\mathcal{H}$, and it follows that $\mathcal{H}$ induces a 3-$(\frac{\tilde{q}^{d-1}-1}{\tilde{q}-1}, k, 1)$ design on which G containing $PSL(d-1, \tilde{q})$ operates as simple normal subgroup. Inductively, we obtain the minimal counter-example for $d = 3$. But, as we have shown above, G with $PSL(3, \tilde{q})$ as simple normal subgroup cannot act on any non-trivial 3-$(\tilde{q}^2 + \tilde{q} + 1, k, 1)$ design.

Case (3): $N = PSU(3, q^2)$, $v = q^3 + 1$, $q = p^e > 2$.

Here $\text{Aut}(N) = P\Gamma U(3, q^2)$, and $|G| = (q^3 + 1)q^3 \frac{(q^2-1)}{n} a$ with $n = (3, q+1)$ and $a \mid 2ne$. Thus, from Remark 4.15, we obtain

$$q^2 + q + 1 = (k - 1)(k - 2)\frac{q+1}{n}\frac{a}{|G_{xB}|} \quad \text{if } x \in B. \tag{7.4}$$

We will show by contradiction that $G \leq \text{Aut}(\mathcal{D})$ cannot act flag-transitively on any non-trivial Steiner 3-design $\mathcal{D}$.

Let $\{v_1, v_2, v_3\}$ be a basis of the non-degenerate hermitian vector space $V = V(3, q^2)$ with

$$(v_2, v_2) = (v_1, v_3) = 1, \ (v_1, v_1) = (v_3, v_3) = (v_1, v_2) = (v_2, v_3) = 0.$$

For $v = \sum_{i=1}^{3} a_i v_i$ and $w = \sum_{i=1}^{3} b_i v_i$ (a_i, $b_i \in GF(q^2)$), we have then

$$(v, w) = a_1 b_3^\tau + a_2 b_2^\tau + a_3 b_1^\tau,$$

where τ denotes the unique involutory automorphism $GF(q^2) \longrightarrow GF(q^2)$, $x \mapsto x^q$. We deduce from [73, Chap. II, Thm. 10.12] that the cyclic group

$$\left\{ \begin{pmatrix} c & & \\ & c^{-2} & \\ & & c \end{pmatrix} \Bigg| \ c \in GF(q^2)^*, \ c^{\tau+1} = 1 \right\}$$

of linear transformations on V induces a group U of dilatations of order $\frac{q+1}{n}$ on the associated projective space $PG(2, q^2)$ with axis the non-absolute line $\mathcal{G}$ consisting of the absolute points $\langle(1, 0, 0)\rangle$, $\langle(0, 0, 1)\rangle$ and $\langle(a_1, 0, a_3)\rangle$ with

$$a_1 a_3^\tau + a_1^\tau a_3 = \text{Tr}(a_1 a_3^\tau) = 0$$

(where Tr denotes the trace map $GF(q^2) \longrightarrow GF(q)$, $x \mapsto x + x^q$) and as center the pole of the axis, i.e., the non-absolute point $\langle(0, 1, 0)\rangle$.

As customary (see, e.g., [6, p. 87]), we call the following non-absolute lines $\mathcal{G}$ and $\mathcal{H}$ *perpendicular* if $\mathcal{G}$ passes through the pole of $\mathcal{H}$ and $\mathcal{H}$ passes, therefore, through the pole of $\mathcal{G}$.

By the definition of Steiner 3-designs, we may choose three distinct absolute points on $\mathcal{G}$, which are incident with a unique block $B \in \mathcal{B}$. Let us first assume that B contains absolute points outside $\mathcal{G}$ which are all on $\mathcal{H}$. It is clear that U fixes each point of $\mathcal{G}$, and hence in particular B. Furthermore, $\mathcal{H}$ intersects $\mathcal{G}$ in a non-absolute point x (see, e.g., [6, p. 88]). As U acts outside x semi-regularly on $\mathcal{H}$, we conclude that all point-orbits have length $\frac{q+1}{n}$. If we choose now three distinct absolute points on $\mathcal{H}$, then they are also incident with the unique block B. Thus, by the same arguments, U fixes each point of $\mathcal{H}$ and acts outside x semi-regularly on $\mathcal{G}$. Therefore, we have

$$k = (n_1 + n_2)\frac{q+1}{n}$$

with $n_1, n_2 \in \{1, 2, 3\}$. If $n = 1$, then obviously $k = 2(q+1)$, which is impossible in view of Lemma 1.14 (c). Thus, $n \neq 1$. For $n_1 + n_2 = 3$, it follows from equation (7.4) that $q^2 + q + 1 \mid (q-1)\frac{a}{n} < q^2 - q$, which is clearly not possible. In each of the other cases, polynomial division with remainder gives a contradiction to Lemma 1.14 (c).

Now, we assume that B contains absolute points outside $\mathcal{G}$ which are not all on $\mathcal{H}$. By applying the same arguments as above, we obtain additionally a lattice of points such that

$$k = n_1 n_2 \left(\frac{q+1}{n}\right)^2 + (n_1 + n_2)\frac{q+1}{n}$$

with n_1, n_2 as above, which clearly contradicts Corollary 1.17.

Hence, we have shown that B is completely contained in $\mathcal{G}$. Thus, in view of the flag-transitivity of G, each block is contained in a non-absolute line. But, by the definition of Steiner 3-designs, any three non-collinear absolute points must also be incident with a unique block, a contradiction.

Case (4): $N = Sz(q)$, $v = q^2 + 1$, $q = 2^{2e+1} > 2$.

We have $\mathrm{Aut}(N) = Sz(q) \rtimes \langle\alpha\rangle$, where α denotes the Frobenius automorphism $GF(q) \longrightarrow GF(q)$, $x \mapsto x^2$. Thus, by Dedekind's law, $G = Sz(q) \rtimes (G \cap \langle\alpha\rangle)$, and $|G| = (q^2 + 1)q^2(q-1)a$ with $a \mid 2e+1$. It follows from Remark 4.15 that

$$q + 1 = (k-1)(k-2)\frac{a}{|G_{xB}|} \quad \text{if } x \in B.$$

We will prove by contradiction that $G \leq \mathrm{Aut}(\mathcal{D})$ cannot act flag-transitively on any non-trivial Steiner 3-design $\mathcal{D}$.

Let us first remark that we only have one conjugacy class of involutions in G. Hence, every involution has exactly one fixed point, which lies in an appropriate block. Therefore, by the flag-transitivity of G, there exists for every $B \in \mathcal{B}$ always an involution $\tau \in G_{xB} \cap Sz(q)$ with $x \in B$, and B can be regarded as the orbit of fixed points of involutions in $G_B \cap Sz(q)$.

Since G is block-transitive, we can restrict ourselves to consider the unique block $B \in \mathcal{B}$ which is incident with the 3-subset $\{0, 1, \infty\}$ of X. As every non-identity element of $Sz(q)$ fixes at most two distinct points, we have $\mathrm{Aut}(N)_{0,1,\infty} = \langle \alpha \rangle$, and thus $G \cap \langle \alpha \rangle \leq G_{0B}$ by the definition of Steiner 3-designs. Setting $u = \frac{|G_{0B}|}{a}$, we next show that $u = 2$ or 4. For the list of subgroups of $Sz(q)$, we refer to [118, Thm. 9]. First, let $G_B \cap Sz(q)$ be isomorphic to $Sz(\bar{q})$ for some $\bar{q} \geq 8$ such that $\bar{q}^m = q$, $m \geq 1$. As B can be regarded as the orbit of fixed points of involutions in $G_B \cap Sz(q)$, it follows that $k = \bar{q}^2 + 1$. Clearly, $m > 1$ (otherwise $k = q^2 + 1$, a contradiction to Corollary 1.17). Thus, we have

$$q + 1 = \bar{q}^2(\bar{q}^2 - 1)\frac{a}{|G_{0B}|}.$$

As $q > 8$, Zsigmondy's Theorem yields the existence of a 2-primitive prime divisor z with $z \perp 2^{2(2e+1)} - 1$. Then

$$z \mid q + 1 = \bar{q}^2(\bar{q}^2 - 1)\frac{a}{|G_{0B}|}.$$

But now Proposition 3.4 gives $(z, \bar{q}) = 1$ and $z > a$ since $z \equiv 1 \pmod{(2e+1)}$. Therefore, we conclude that $\bar{q} = q$, a contradiction.

Let $G_B \cap Sz(q)$ be conjugate to a subgroup of $Sz(q)_x$ $(x \in X)$. By the transitivity of G, we can choose x as fixed point of an involution. Thus, $x \in B$ by the remark above, contrary to the fact that $x \notin B$ by the flag-transitivity of G.

Let $G_B \cap Sz(q)$ be conjugate to a subgroup of U with $|U| = 4(q \pm l + 1)$, where $l^2 = 2q$. Then $|O_{p'}(U)| = q \pm l + 1$, and $O_{p'}(U)$ operates fixed-point-freely on X since $(q \pm l + 1, q) = 1$ and $(q \pm l + 1, q^2 - 1) = 1$. Thus $(G_{0B} \cap Sz(q)) \cap O_{p'}(U) = 1$, and therefore $|G_{0B} \cap Sz(q)| \leq 4$.

Let $G_B \cap Sz(q)$ be conjugate to a subgroup of U with $|U| = 2(q - 1)$. Then $|O_{p'}(U)| = q - 1$, and $O_{p'}(U)$ has two distinct fixed points in X. As $O_{p'}(U)$ contains no involutions, these fixed points cannot lie in B by the remark above. Hence $(G_{0B} \cap Sz(q)) \cap O_{p'}(U) = 1$, and thus $|G_{0B} \cap Sz(q)| \leq 2$. Since $|G_{0B} \cap Sz(q)| \equiv 0 \pmod 2$, we have therefore

$$|G_{0B} \cap Sz(q)| = 2 \text{ or } 4.$$

As $G \cap \langle \alpha \rangle \leq G_{0B}$, and clearly $(G_B \cap Sz(q)) \cap (G \cap \langle \alpha \rangle) = 1$, we conclude that

$$u = 2 \text{ or } 4.$$

Finally, our equation

$$u(q + 1) = (k - 1)(k - 2)$$

implies for $u = 2$ that

$$2^{2e+2} = k(k - 3),$$

which is clearly impossible since $e \geq 1$. For $u = 4$, we obtain the Thue-Mahler equation

$$2^{2e+3} = k^2 - 3k - 2. \tag{7.5}$$

By setting $x = 2k - 3$ and $n = 2e + 5$ this becomes the well-known generalized Ramanujan-Nagell equation

$$x^2 - 17 = 2^n,$$

which has exactly the four solutions $(x, n) = (5, 3), (7, 5), (9, 6), (23, 9)$ (see, e.g., [10, Thm. 3]). As we have $e \geq 1$, it follows that $(e, k) = (2, 13)$ is the only solution of equation (7.5). But, by Lemma 1.14 (b), this is impossible, which verifies the claim.

Case (5): $N = Re(q)$, $v = q^3 + 1$, $q = 3^{2e+1} > 3$.

Here $\mathrm{Aut}(N) = Re(q) \rtimes \langle \alpha \rangle$, where α denotes the Frobenius automorphism $GF(q) \longrightarrow GF(q)$, $x \mapsto x^3$. Thus, by Dedekind's law, $G = Re(q) \rtimes (G \cap \langle \alpha \rangle)$, and $|G| = (q^3 + 1)q^3(q - 1)a$ with $a \mid 2e + 1$. From Remark 4.15, we hence obtain

$$q^2 + q + 1 = (k - 1)(k - 2)\frac{a}{|G_{xB}|} \quad \text{if } x \in B. \tag{7.6}$$

We will also prove by contradiction that $G \leq \mathrm{Aut}(\mathcal{D})$ cannot act flag-transitively on any non-trivial Steiner 3-design $\mathcal{D}$.

We remark that we only have one conjugacy class of involutions in G. Thus, every involution fixes at least three distinct points, each of which lies in an appropriate block. Therefore, by the flag-transitivity of G, there exists for every $B \in \mathcal{B}$ always an involution $\tau \in G_{xB} \cap Re(q)$ with $x \in B$.

We show furthermore that 9 divides the order of $G_B \cap Re(q)$. Let P be a Sylow 3-subgroup of $Re(q)$. According to [122], P contains a normal elementary Abelian subgroup $\overline{P}$ of order q^2 containing $Z(P)$. Thus, there exist subgroups U_1, U_2 of $\overline{P}$ of order 3 with $U_1 \leq Z(P)$, $U_2 \not\leq Z(P)$. As the stabilizer of three distinct points in $Re(q)$ has order 2, we have $\mathrm{Fix}_X(U_1) = \mathrm{Fix}_X(U_2) = \{x\}$ for some $x \in X$. Hence, if U_1 and U_2 are conjugate in $Re(q)$, then they are already conjugate in $Re(q)_x$. But, as $Z(P)$ is a characteristic subgroup of $Re(q)_x$, this is impossible. Therefore, we have at least two distinct conjugacy classes of subgroups of order 3 in $Re(q)$, and the assertion follows by the definition of Steiner 3-designs.

Because of the block-transitivity of G, we can restrict ourselves to consider the unique block $B \in \mathcal{B}$ which is incident with the 3-subset $\{0, 1, \infty\}$ of X. Clearly, $\langle \alpha \rangle \leq \mathrm{Aut}(N)_{0,1,\infty}$, and hence $G \cap \langle \alpha \rangle \leq G_{0B}$ by the definition of Steiner 3-designs. Furthermore, obviously $(G_B \cap Re(q)) \cap (G \cap \langle \alpha \rangle) = 1$. Therefore, as G_B acts transitively on the points of B, Dedekind's law yields

$$k = \left|0^{G_B}\right| = [G_B : G_{0B}] = [G_B \cap Re(q) : G_{0B} \cap Re(q)]. \tag{7.7}$$

Thus, $G_B \cap Re(q)$ acts also transitively on the points of B.

In the following, we will examine the list of subgroups of $Re(q)$ (cf. [122]). As 9 divides the order of $G_B \cap Re(q)$, clearly $G_B \cap Re(q)$ cannot be conjugate to a subgroup of the normalizer of a Sylow 2-subgroup of $Re(q)$ of order $8 \cdot 7 \cdot 3$. By the same argument, $G_B \cap Re(q)$ cannot be conjugate to a subgroup of U with $|U| = 6(q + 1 \pm 3l)$, where $l = 3^e$.

Let $G_B \cap Re(q)$ be isomorphic to $Re(\bar{q})$ for some $\bar{q} \geq 27$ such that $\bar{q}^m = q$, $m \geq 1$. Let $\overline{X} \subseteq X$ with $|\overline{X}| = \bar{q}^3 + 1$. We first show that only involutions may have fixed points in $X \setminus \overline{X}$. Let $g \in G$ with $o(g) = s$, where $s \neq 2$ is a prime. If $s \mid \bar{q} - 1$, then g has two distinct fixed points in $\overline{X}$, and none in $X \setminus \overline{X}$, since the stabilizer of three distinct points in $Re(q)$ has order 2. For $s = 3$, clearly g has exactly one fixed point, which lies in $\overline{X}$. If $s \mid \bar{q} + 1$, we show that g has no fixed point in X. Obviously, g has no fixed point in $\overline{X}$. As $3 \nmid \bar{q} + 1$, we assume that g has two distinct fixed points in $X \setminus \overline{X}$. But, as

$$q^3 - \bar{q}^3 = \left(\sum_{i=0}^{3m-1} (-1)^i \frac{q^3}{\bar{q}^{i+1}} - \bar{q}^2 + \bar{q} - 1 \right)(\bar{q} + 1),$$

and hence $(q^3 - \bar{q}^3 - 2, \bar{q} + 1) = (2, \bar{q} + 1) = 2$, this is impossible. If $s \mid \bar{q} + 1 \pm 3\bar{l}$, we show again that g has no fixed point in X. As $\bar{q}^3 + 1 = (\bar{q} + 1 + 3\bar{l})(\bar{q} + 1 - 3\bar{l})(\bar{q} + 1)$, it is obvious that g has no fixed point in $\overline{X}$. Since $3 \nmid \bar{q} + 1 \pm 3\bar{l}$, we assume in both cases that g has two distinct fixed points in $X \setminus \overline{X}$. But, as $(\bar{q} + 1 + 3\bar{l})(\bar{q} + 1 - 3\bar{l}) = \bar{q}^2 - \bar{q} + 1$, and

$$q^3 - \bar{q}^3 = \left(\sum_{i=0}^{m-1} \sum_{j=0}^{1} (-1)^{2+3i} \frac{q^3}{\bar{q}^{2+3i+j}} - \bar{q} - 1 \right)(\bar{q}^2 - \bar{q} + 1),$$

we have $(q^3 - \bar{q}^3 - 2, \bar{q}^2 - \bar{q} + 1) = (2, \bar{q}^2 - \bar{q} + 1) = 2$, a contradiction.

As $G_B \cap Re(q)$ acts transitively on the points of B, we have $B \subseteq \overline{X}$ or $B \subseteq X \setminus \overline{X}$. In the first case, equation (7.7) gives

$$k = \bar{q}^3 + 1,$$

while in the second

$$k = \frac{(\bar{q}^3 + 1)\bar{q}^3(\bar{q} - 1)}{n},$$

where n is a power of 2, and $n \leq 8$ as the order of $Re(q)$ is divisible by 8 but not by 16.

We will prove now that none of these values of k is possible. We assume first that $k = \bar{q}^3 + 1$. Clearly, $m > 1$ (otherwise $k = q^3 + 1$, a contradiction to Corollary 1.17). Thus, we have

$$q^2 + q + 1 = \bar{q}^3(\bar{q}^3 - 1)\frac{a}{|G_{0B}|}.$$

Zsigmondy's Theorem yields the existence of a 3-primitive prime divisor z with $z \perp 3^{3(2e+1)} - 1$. Then

$$z \mid q^2 + q + 1 = \bar{q}^3(\bar{q}^3 - 1)\frac{a}{|G_{0B}|}.$$

But now Proposition 3.4 yields $(z, \bar{q}) = 1$ and $z > a$ since $z \equiv 1 \pmod{(2e+1)}$. Therefore, we have $\bar{q} = q$, a contradiction.

Now, we assume that $k = \frac{(\bar{q}^3+1)\bar{q}^3(\bar{q}-1)}{n}$. Then

$$|G_{0B}| = \frac{|G_B \cap Re(q)|\,\bar{a}}{k} = n\bar{a},$$

where $\bar{a} \mid a$. Here, $n < 4$ since otherwise $(k-1)(k-2) \equiv 0 \pmod 4$ by equation (7.6) and, by applying Lemma 1.14 (c), this would imply that $q^3 - 1$ is divisible by 4, which is impossible since $q - 1 \equiv 2 \pmod 8$ in $Re(q)$. Thus, we may assume that $n = 2$. Polynomial division with remainder gives

$$q^3 - 1 = \left(\sum_{i=0}^{\overline{m}} \frac{2^{2i+1}q^3}{\left((\bar{q}^3+1)\bar{q}^3(\bar{q}-1)\right)^{i+1}} \right) \left(\frac{(\bar{q}^3+1)\bar{q}^3(\bar{q}-1)}{2} - 2 \right)$$
$$+ \frac{2^{2\overline{m}+2}q^3}{\left((\bar{q}^3+1)\bar{q}^3(\bar{q}-1)\right)^{\overline{m}+1}} - 1$$

for a suitable $\overline{m} \in \mathbb{N}$ (such that

$$\deg\left(\frac{2^{2\overline{m}+2}q^3}{\left((\bar{q}^3+1)\bar{q}^3(\bar{q}-1)\right)^{\overline{m}+1}} - 1 \right) < \deg\left(\frac{(\bar{q}^3+1)\bar{q}^3(\bar{q}-1)}{2} - 2 \right)$$

as is well-known). As $8 \mid |Re(\bar{q})|$, clearly $\left((\bar{q}^3+1)\bar{q}^3(\bar{q}-1)\right)^{\overline{m}+1}$ is divisible by $2^{3(\overline{m}+1)}$. Thus $\frac{2^{2\overline{m}+2}q^3}{\left((\bar{q}^3+1)\bar{q}^3(\bar{q}-1)\right)^{\overline{m}+1}} \neq 1$, yielding a contradiction to Lemma 1.14 (c).

Let $G_B \cap Re(q)$ be conjugate to a subgroup of $Re(q)_x$ ($x \in X$). By the transitivity of G, we can choose x as fixed point of an involution. Thus, $x \in B$ for an appropriate block $B \in \mathcal{B}$ by the remark above, contrary to the fact that $x \notin B$ by the flag-transitivity of G.

Let $G_B \cap Re(q)$ be conjugate to a subgroup of $PSL(2, q) \times \langle \tau \rangle$, where τ denotes any involution in $Re(q)$. By the remark above, we can choose τ such that 0 is a fixed point under τ. As 9 must be a divisor of the order of $G_B \cap Re(q)$, we can restrict ourselves to the examination of the following cases (cf. [44, Chap. 12, p. 285f.] or [73, Chap. II, Thm. 8.27]):

(i) $G_B \cap Re(q)$ is conjugate to $PSL(2, \bar{q})$ or $PSL(2, \bar{q}) \times \langle \tau \rangle$ for some $\bar{q} \geq 27$ such that $\bar{q}^m = q$, $m \geq 1$.

Let $\overline{X} \subseteq X$ with $|\overline{X}| = \bar{q}+1$. First, we show again that only involutions may have fixed points in $X \setminus \overline{X}$. Let $g \in G$ with $o(g) = s$, where $s \neq 2$ is a prime. If $s \mid \bar{q} - 1$, then g has two distinct fixed points in $\overline{X}$ and none in $X \setminus \overline{X}$. For $s = 3$, clearly g has exactly one fixed point, which lies in $\overline{X}$. If $s \mid \bar{q} + 1$, we show that g has no fixed point in X. Obviously, g has no fixed point in $\overline{X}$. As $3 \nmid \overline{q} + 1$, we assume that g has two distinct fixed points in $X \setminus \overline{X}$. But, as

$$q^3 - \overline{q} = \left(\sum_{i=0}^{\overline{3m}-1} (-1)^i \frac{q^3}{\overline{q}^{i+1}} - 1 \right) (\overline{q} + 1),$$

and hence $(q^3 - \overline{q} - 2, \overline{q} + 1) = (2, \overline{q} + 1) = 2$, this is impossible.

Again, we have $B \subseteq \overline{X}$ or $B \subseteq X \setminus \overline{X}$. With equation (7.7), we obtain

$$k = \overline{q} + 1,$$

in the first case, while in the second

$$k = \frac{\overline{q}(\overline{q}^2 - 1)}{n},$$

where n is a power of 2, and $n \le 8$ again.

We will prove now that none of the values of k is possible. We assume first that $k = \overline{q} + 1$. Then

$$q^2 + q + 1 \mid \overline{q}(\overline{q} - 1)a$$

by equation (7.6). Since $(q^2+q+1, \overline{q}) = 1$ and $(q^2+q+1, \overline{q}-1) = (3, \overline{q}-1) = 1$, this is equivalent to

$$q^2 + q + 1 \mid a,$$

which is impossible as clearly $a \le q$. Now, we assume that $k = \frac{\overline{q}(\overline{q}^2-1)}{n}$. Then

$$|G_{0B}| = \frac{|G_B \cap Re(q)|\overline{a}}{k} = \frac{n\overline{a}}{2} \text{ or } n\overline{a},$$

where $\overline{a} \mid a$. Considering the first gives

$$(q^2 + q + 1)\frac{n}{2} = (k - 1)(k - 2)\frac{a}{\overline{a}}$$

by equation (7.6). Clearly, $n = 2$ is impossible. If $n = 4$, then $k = \frac{\overline{q}(\overline{q}^2-1)}{4}$ is divisible by 2 but not by 4. Thus, 4 is a divisor of $k - 2$, but not of the left-hand side. For $n = 8$, we have $(k - 1)(k - 2) \equiv 0 \pmod 4$, which is not possible as we have seen above.

Now, we assume that $|G_{0B}| = n\overline{a}$. Here, $n < 4$ again. For $n = 2$, we have $k = \frac{\overline{q}^3 - \overline{q}}{2}$. Then, polynomial division with remainder gives

$$q^3 - 1 = \left(\sum_{i=0}^{\overline{m}} \frac{2^{2i+1}q^3}{(\overline{q}^3 - \overline{q})^{i+1}} \right) \left(\frac{\overline{q}^3 - \overline{q}}{2} - 2 \right) + \frac{2^{2\overline{m}+2}q^3}{(\overline{q}^3 - \overline{q})^{\overline{m}+1}} - 1$$

for a suitable $\overline{m} \in \mathbb{N}$. As $(\overline{q}^2-1)^{\overline{m}+1}$ is divisible by $2^{3(\overline{m}+1)}$, clearly $\frac{2^{2\overline{m}+2}q^3}{(\overline{q}^3-\overline{q})^{\overline{m}+1}} \ne 1$, yielding a contradiction to Lemma 1.14 (c).

(ii) $G_B \cap Re(q)$ is conjugate to U or $U \times \langle \tau \rangle$, where U is an elementary Abelian subgroup of order $\bar{q} \mid q$ of $PSL(2, q)$.

Let $\overline{X} \subseteq X$ with $|\overline{X}| = q+1$. Clearly, U operates regularly on $\bar{q}$ points, and each non-identity element of U has ∞ as its only fixed point in $\overline{X}$ and none in $X \setminus \overline{X}$. As $2 \nmid |U|$, it follows that $k = \bar{q}$ in both of the cases $B \subseteq \overline{X}$ and $B \subseteq X \setminus \overline{X}$. But, polynomial division with remainder gives

$$q^3 - 1 = \left(\sum_{i=0}^{\overline{m}} \frac{2^i q^3}{\bar{q}^{i+1}} \right) \left(\bar{q} - 2 \right) + \frac{2^{\overline{m}+1} q^3}{\bar{q}^{\overline{m}+1}} - 1$$

for a suitable $\overline{m} \in \mathbb{N}$. As clearly $\frac{2^{\overline{m}+1} q^3}{\bar{q}^{\overline{m}+1}} \neq 1$, this again leads to a contradiction to Lemma 1.14 (c).

(iii) $G_B \cap Re(q)$ is conjugate to U or $U \times \langle \tau \rangle$, where U is a semi-direct product of an elementary Abelian subgroup of order $\bar{q} \mid q$ with a cyclic subgroup of order c of $PSL(2, q)$ with $c \mid \bar{q} - 1$ and $c \mid q - 1$.

Let $\overline{X} \subseteq X$ with $|\overline{X}| = q+1$. Again, we show that only involutions may have fixed points in $X \setminus \overline{X}$. Let $g \in G$ with $o(g) = s$, where $s \neq 2$ is a prime. If $s = 3$, then g has exactly one fixed point, which lies in $\overline{X}$. If $s \mid c$, then g has exactly two distinct fixed points, which lie in $\overline{X}$.

For $B \subseteq \overline{X}$, we deduce that $k = \bar{q}$ or $\bar{q}c$, and for $B \subseteq X \setminus \overline{X}$ that $k = \frac{\bar{q}c}{n}$ with $n \leq 2$ since $q - 1 \equiv 2 \pmod 8$. Again, we will prove that none of the values of k is possible. For $k = \bar{q}$, we have already shown that this is impossible. We assume next that $k = \bar{q}c$. If $2 \mid c$, then k is divisible by 2 but not by 4. Therefore, $k - 2 \equiv 0 \pmod 4$, and hence $q^3 - 1 \equiv 0 \pmod 4$ by Lemma 1.14 (c), which is impossible as we have already seen. For $2 \nmid c$, polynomial division with remainder gives

$$q^3 - 1 = \left(\sum_{i=0}^{\overline{m}} \frac{2^i q^3}{(\bar{q}c)^{i+1}} \right) \left(\bar{q}c - 2 \right) + \frac{2^{\overline{m}+1} q^3}{(\bar{q}c)^{\overline{m}+1}} - 1$$

for a suitable $\overline{m} \in \mathbb{N}$. But obviously $\frac{2^{\overline{m}+1} q^3}{(\bar{q}c)^{\overline{m}+1}} \neq 1$, which leads to the same contradiction as before.

Now, we assume that $k = \frac{\bar{q}c}{n}$. Then

$$|G_{0B}| = \frac{|G_B \cap Re(q)| \, \bar{a}}{k} = n\bar{a} \text{ or } 2n\bar{a},$$

where $\bar{a} \mid a$. When considering the first possibility, clearly equation (7.6) rules out the case $n = 1$. So, we assume that $n = 2$. Hence $k = \frac{\bar{q}c}{2}$, but polynomial division with remainder gives

$$q^3 - 1 = \left(\sum_{i=0}^{\overline{m}} \frac{2^{2i+1} q^3}{(\bar{q}c)^{i+1}} \right) \left(\frac{\bar{q}c}{2} - 2 \right) + \frac{2^{2\overline{m}+2} q^3}{(\bar{q}c)^{\overline{m}+1}} - 1$$

for a suitable $\overline{m} \in \mathbb{N}$. But since $c \mid q - 1$, the largest possible power of 2 that is contained in $c^{\overline{m}+1}$ is $2^{\overline{m}+1}$. Thus $\frac{2^{2\overline{m}+2}q^3}{(\overline{q}c)^{\overline{m}+1}} \neq 1$, the same contradiction as above.

Now, we assume that $|G_{0B}| = 2n\overline{a}$. For $n = 1$, we get $k = \overline{q}c$, which is not possible as shown above. The case $n = 2$ is ruled out by equation (7.6) since $(k-1)(k-2)$ is not divisible by 4 as we already know.

This completes the list of subgroups of $Re(q)$ that we have to examine, and the claim is established.

Case (6): $N = Sp(2d, 2)$, $d \geq 3$, $v = 2^{2d-1} \pm 2^{d-1}$.

As here $|\mathrm{Out}(N)| = 1$, we have $N = G$. Let X^+ respectively X^- denote the set of points on which G operates. It is well-known that G_x acts on $X^\pm \setminus \{x\}$ as $O^\pm(2d, 2)$ does in its usual rank 3 representation on singular points of the underlying orthogonal space. Thus, G_{xy} has two orbits on $X^\pm \setminus \{x, y\}$ of length $2(2^{d-1}\mp1)(2^{d-2}\pm1)$ and 2^{2d-2} (see, e.g., [78, p. 69]). We will show by contradiction that $G \leq \mathrm{Aut}(\mathcal{D})$ cannot act flag-transitively on any non-trivial Steiner 3-design $\mathcal{D}$.

Let $z \in X^\pm \setminus \{x, y\}$. Then, in both cases, the 3-subset $\{x, y, z\}$ is incident with a unique block $B \in \mathcal{B}$. By Remark 4.15, we have therefore

$$(v - 2)\,|G_{xB}| = (k - 1)(k - 2)\,|G_{xy}|, \tag{7.8}$$

where

$$|G_{xB}| = n\frac{|G_{xy}|}{|z^{G_{xy}}|}$$

for some $n \in \mathbb{N}$. This is equivalent to

$$2(2^{2d-2} \pm 2^{d-2} - 1)n = (k - 1)(k - 2)\left|z^{G_{xy}}\right|$$

with

$$\left|z^{G_{xy}}\right| = \begin{cases} 2(2^{d-1} \mp 1)(2^{d-2} \pm 1), & \text{or} \\ 2^{2d-2}. \end{cases}$$

Clearly, $2^{2d-2} \pm 2^{d-2} - 1 \equiv 1 \pmod 2$ and $(k - 1)(k - 2) \equiv 0 \pmod 2$. As $(2^{2d-2} \pm 2^{d-2} - 1, 2^{d-1} \mp 1) = (2^{d-2}, 2^{d-1} \mp 1) = 1$ and $(2^{2d-2} \pm 2^{d-2} - 1, 2^{d-2} \pm 1) = (2, 2^{d-2} \pm 1) = 1$, it follows that $\left|z^{G_{xy}}\right|$ always divides n. Thus $|G_{xy}| \mid |G_{xB}|$, and equation (7.8) implies

$$v - 2 \mid (k - 1)(k - 2).$$

But, on the other hand, we have $v - 2 \geq (k - 1)(k - 2)$ by Proposition 1.16 (b), and it is immediately seen that v cannot take the values where equality holds.

Case (9): $N = M_v$, $v = 11, 12, 22, 23, 24$.

Here G is always 3-transitive, and thus Theorem 5.2 gives the design described in Part (iv) of Theorem 7.1. Obviously, flag-transitivity holds as the 3-transitivity of G implies that G_x acts block-transitively on the derived Steiner 2-design $\mathcal{D}_x$ for any $x \in X$.

This completes the proof of Theorem 7.1.   □

Remark 7.6. Referring to the generalized Ramanujan-Nagell equation in Case (4) of this section, we note that S. Ramanujan [109] conjectured in 1913 that the Diophantine equation of second order

$$x^2 + 7 = 2^n$$

in the positive integers x and n only has exactly the five solutions

$$(x, n) = (1, 3), (3, 4), (5, 5), (11, 7), (181, 15)$$

(see, e.g., [109, p. 327]). This was first verified by T. Nagell [105] in 1948. Various generalizations of this Ramanujan-Nagell equation have been considered by, e.g., H. Hasse [53] in 1966 and F. Beukers [10, 11] in 1980/81.

Chapter 8

The Classification of Flag-transitive Steiner 4-Designs

8.1 Introduction

We describe the complete classification of all flag-transitive Steiner 4-designs in this chapter. Our approach again uses the classification of the finite doubly transitive permutation groups. For the groups of affine type, incidence geometric arguments can be applied similarly as in the previous chapter. For the groups of almost simple type, surprisingly with the projective group $PSL(2,q)$ as simple normal subgroup – although group-theoretically very well understood – sophisticated arguments seem to be necessary. In this regard we will first determine the orbit-lengths from the action of subgroups of $PSL(2,q)$ on the points of the projective line (Lemmas 8.3–8.11).

8.2 Main Result

The classification of all non-trivial Steiner 4-designs admitting a flag-transitive group of automorphisms is as follows:

Theorem 8.1. *Let $\mathcal{D}=(X,\mathcal{B},I)$ be a non-trivial Steiner 4-design. Then $G \leq \mathrm{Aut}(\mathcal{D})$ acts flag-transitively on $\mathcal{D}$ if and only if one of the following occurs:*

(1) *$\mathcal{D}$ is isomorphic to the Mathieu-Witt 4-$(11,5,1)$ design, and $G \cong M_{11}$,*

(2) *$\mathcal{D}$ is isomorphic to the Mathieu-Witt 4-$(23,7,1)$ design, and $G \cong M_{23}$.*

8.3   Groups of Automorphisms of Affine Type

In the sequel, we start with the proof of Theorem 8.1. Let $\mathcal{D} = (X, \mathcal{B}, I)$ be a non-trivial Steiner 4-design with $G \leq \mathrm{Aut}(\mathcal{D})$ acting flag-transitively on $\mathcal{D}$ throughout this chapter. We recall that due to Proposition 4.13, we may restrict ourselves to examination of the finite 2-transitive permutation groups. Clearly, in the following we may assume that $k > 4$ as trivial Steiner 4-designs are excluded. Let us assume in this section that G is of affine type. Each of the Cases (5)–(8) can easily be ruled out by hand using Lemma 1.14 (b) and (c), Corollary 1.17, and Lemma 4.1.

Case (1): $G \leq A\Gamma L(1, v)$, $v = p^d$.

As G is point 2-transitive, we have $|G| = v(v - 1)a$ with $a \mid d$. Using Lemma 4.1, we obtain

$$(p^d - 2)(p^d - 3) \mid a(k - 1)(k - 2)(k - 3) \mid d(k - 1)(k - 2)(k - 3),$$

and hence in particular

$$(p^d - 2)(p^d - 3) \leq d(k - 1)(k - 2)(k - 3).$$

But, Proposition 1.16 (b) gives

$$p^d - 3 \geq (k - 2)(k - 3),$$

and thus

$$p^d - 2 \leq d(k - 1).$$

With regard to Corollary 1.17, this leaves only a very small number of possibilities for k to check, which can easily be ruled out by hand using Lemma 1.14 (b) and (c). Therefore, $G \leq \mathrm{Aut}(\mathcal{D})$ cannot act flag-transitively on any non-trivial Steiner 4-design $\mathcal{D}$.

Case (2): $G_0 \trianglerighteq SL(\frac{d}{a}, p^a)$, $d \geq 2a$.

Let e_i denote the i-th standard basis vector of the vector space $V = V(\frac{d}{a}, p^a)$, and $\langle e_i \rangle$ the 1-dimensional vector subspace spanned by e_i.

First, let $p^a > 3$. For $d = 2a$, let $U = U(\langle e_1 \rangle) \leq G_0$ denote the subgroup of all transvections with axis $\langle e_1 \rangle$. Clearly, U fixes as points only the elements of $\langle e_1 \rangle$. Thus, G_0 has point-orbits of length at least p^a outside $\langle e_1 \rangle$. Let $S = \{0, e_1, x, y\}$ be a 4-subset of distinct points with $x, y \in \langle e_1 \rangle$. Obviously, U fixes the unique block $B \in \mathcal{B}$ which is incident with S. If B contains at least one point outside $\langle e_1 \rangle$, then we would obtain $k \geq p^a + 4$, which is not possible as $k \leq p^a + 2$ in view of Corollary 1.17. Hence B is contained completely in $\langle e_1 \rangle$, and as G is flag-transitive, we conclude that each block lies in an affine line. But, by the definition of Steiner 4-designs, any four distinct non-collinear points must also be incident with a unique block, a contradiction.

For $d \geq 3a$, $SL(\frac{d}{a}, p^a)_{e_1}$ and hence also G_{0, e_1} acts point-transitively on $V \setminus \langle e_1 \rangle$. Again, let $S = \{0, e_1, x, y\}$ be a 4-subset of distinct points with $x, y \in \langle e_1 \rangle$.

If the unique block $B \in \mathcal{B}$ which is incident with S contains some point outside $\langle e_1 \rangle$, then it would already contain all points outside, thus at least $p^d - p^a + 4$ many, which obviously contradicts Corollary 1.17. We conclude that B lies completely in $\langle e_1 \rangle$, and thus may proceed as above.

Now, let $p^a = 2$. Then $v = 2^d$. For $d = 3$, we have $v = 8$ and $k = 5$ by Corollary 1.17, which is not possible in view of Lemma 1.14 (c). Therefore, we assume that $d > 3$. We remark that clearly any three distinct points are non-collinear in $AG(d, 2)$ and hence define an affine plane. Let $\mathcal{E} = \langle e_1, e_2 \rangle$ denote the 2-dimensional vector subspace spanned by e_1 and e_2. Then $SL(d, 2)_\mathcal{E}$ and hence also $G_{0,\mathcal{E}}$ acts point-transitively on $V \setminus \mathcal{E}$. If the unique block $B \in \mathcal{B}$ which is incident with the 4-subset $\{0, e_1, e_2, e_1 + e_2\}$ contains some point outside $\mathcal{E}$, then it would already contain all points of $V \setminus \mathcal{E}$, and hence $k \geq 2^d - 4 + 4 = 2^d$, a contradiction to Corollary 1.17. Therefore, B can be identified with $\mathcal{E}$, and the flag-transitivity of G implies that each block must be an affine plane, a contradiction as $k > 4$. Similar arguments hold for $p^a = 3$.

Case (3): $G_0 \trianglerighteq Sp(\frac{2d}{a}, p^a)$, $d \geq 2a$.

First, let $p^a \neq 2$. The permutation group $PSp(\frac{2d}{a}, p^a)$ on the points of the associated projective space is a rank 3 group, and the orbits of the one-point stabilizer are well-known (e.g. [73, Chap. II, Thm. 9.15 (b)]). Thus, $G_0 \trianglerighteq Sp(\frac{2d}{a}, p^a)$ has exactly two orbits on $V \setminus \langle x \rangle$ $(0 \neq x \in V)$ of length at least

$$\frac{p^a(p^{2d-2a} - 1)}{p^a - 1} = \sum_{i=1}^{\frac{2d}{a} - 2} p^{ia} > p^d.$$

Let $S = \{0, x, y, z\}$ be a 4-subset with $y, z \in \langle x \rangle$. If the unique block incident with S contains at least one point of $V \setminus \langle x \rangle$, then we would have $k > p^d + 4$, which is impossible since $k \leq p^d + 2$ by Corollary 1.17. Therefore, we can argue as in the previous case.

Now, let $p^a = 2$. Then $v = 2^{2d}$. For $d = 2$ (here $Sp(4, 2) \cong S_6$ as well-known), Corollary 1.17 gives $k \leq 6$. But, Lemma 1.14 (c) rules out the cases when $k = 5$ or 6. Thus, let $d > 2$. It is easily seen that there are $2^{2d-1}(2^{2d} - 1)$ hyperbolic pairs in the non-degenerate symplectic space $V = V(2d, 2)$, and by Witt's theorem, $Sp(2d, 2)$ is transitive on these hyperbolic pairs (cf. [73, Chap. II, Thm. 9.13]). Let $\{x, y\}$ denote a hyperbolic pair, and $\mathcal{E} = \langle x, y \rangle$ the hyperbolic plane spanned by $\{x, y\}$. As $\mathcal{E}$ is non-degenerate, we have the orthogonal decomposition

$$V = \mathcal{E} \perp \mathcal{E}^\perp.$$

Obviously, $Sp(2d, 2)_{\{x,y\}}$ stabilizes $\mathcal{E}^\perp$ as a subspace, which implies that $Sp(2d, 2)_{\{x,y\}} \cong Sp(2d - 2, 2)$. As $\text{Out}(Sp(2d, 2)) = 1$, we have therefore

$$Sp(2d - 2, 2) \cong Sp(2d, 2)_{\{x,y\}} \trianglelefteq Sp(2d, 2)_\mathcal{E} = G_{0,\mathcal{E}}.$$

As $Sp(2d - 2, 2)$ acts transitively on the non-zero vectors of the $(2d - 2)$-dimensional symplectic subspace, the smallest orbit on $V \setminus \mathcal{E}$ under $G_{0,\mathcal{E}}$ has length at least $2^{2d-2} - 1$. If the unique block $B \in \mathcal{B}$ which is incident with the 4-subset $\{0, x, y, x + y\}$ contains some point in $V \setminus \mathcal{E}$, then we would have $k \geq 2^{2d-2} + 3$, a contradiction to Corollary 1.17. Thus B can be identified with $\mathcal{E}$, and by the flag-transitivity of G, each block must be an affine plane, yielding a contradiction.

Case (4): $G_0 \trianglerighteq G_2(2^a)'$, $d = 6a$.

First, let $a = 1$. Then $v = 2^6 = 64$ and $k \leq 10$ by Corollary 1.17. On the other hand, we have $|G_2(2)'| = 2^5 \cdot 3^3 \cdot 7$ and $|\mathrm{Out}(G_2(2)')| = 2$. In view of Lemma 4.1 this gives

$$r = \frac{63 \cdot 62 \cdot 61}{(k-1)(k-2)(k-3)} \,\Big|\, |G_0| \,\Big|\, 2^6 \cdot 3^3 \cdot 7,$$

which implies that k is at least 63, a contradiction.

Now, let $a > 1$. As here $G_2(2^a)$ is simple non-Abelian, it is sufficient to consider $G_0 \trianglerighteq G_2(2^a)$. The permutation group $G_2(2^a)$ is of rank 4, and for $0 \neq x \in V$ the one-point stabilizer $G_2(2^a)_x$ has exactly three orbits $\mathcal{O}_i$ $(i = 1, 2, 3)$ on $V \setminus \langle x \rangle$ of length $2^{3a} - 2^a, 2^{5a} - 2^{3a}, 2^{6a} - 2^{5a}$ (cf., e.g., [2] or [27, Thm. 3.1]). Thus, G_0 has exactly three orbits on $V \setminus \langle x \rangle$ of length at least $|\mathcal{O}_i|$. Let $S = \{0, x, y, z\}$ be a 4-subset with $y, z \in \langle x \rangle$. If the unique block $B \in \mathcal{B}$ which is incident with S contains at least one point of $V \setminus \langle x \rangle$ in $\mathcal{O}_2$ or $\mathcal{O}_3$, then we would obtain as above a contradiction to Corollary 1.17. Thus, we only have to consider the case when B contains points of $V \setminus \langle x \rangle$ which all lie in $\mathcal{O}_1$. By [2], the orbit $\mathcal{O}_1$ is exactly known, and we have

$$\mathcal{O}_1 = x\Delta \setminus \langle x \rangle,$$

where $x\Delta = \{y \in V \mid f(x, y, z) = 0 \text{ for all } z \in V\}$ with an alternating trilinear form f on V. Then B consists, apart from elements of $\langle x \rangle$, exactly of $\mathcal{O}_1$. Since $|\mathcal{O}_1| \neq 1$, we can choose $\langle \overline{x} \rangle \in x\Delta$ with $\langle \overline{x} \rangle \neq \langle x \rangle$. But then, for symmetric reasons, the 4-subset $\{0, \overline{x}, \overline{y}, \overline{z}\}$ with $\overline{y}, \overline{z} \in \langle \overline{x} \rangle$ must also be incident with the unique block B, a contradiction to the fact that $\overline{x}\Delta \neq x\Delta$ for $\langle \overline{x} \rangle \neq \langle x \rangle$. Consequently, B is contained completely in $\langle x \rangle$, and we may argue as in the cases above.

8.4   Groups of Automorphisms of Almost Simple Type

We consider in this section successively those cases where G is of almost simple type. Each of the Cases $(8), (11)$–(13) can easily be eliminated by hand using Lemma 1.14 (b) and (c), Corollary 1.17, and Lemma 4.1. Before we proceed, we prove some lemmas which will be required for Case (2).

In the following, let q be a prime power p^e, and U a subgroup of $PSL(2, q)$. Furthermore, let N_l denote the number of orbits of length l and let $n = (2, q - 1)$. We will determine the orbit-lengths from the action of subgroups of $PSL(2, q)$ on the points of the projective line. Thereby, we remark that for subgroups $U_1 \leq U_2 \leq PSL(2, q)$, any orbit of U_2 is a union of orbits of U_1. For the list of subgroups of $PSL(2, q)$, we refer to [44, Chap. 12, p. 285f.] or [73, Chap. II, Thm. 8.27]. In the special case when $q \equiv 3 \pmod 4$, the orbit lengths have also been calculated in [28, Sect. 4].

Well-known is the following fact (see, e.g., [73, Chap. II, p. 191f.]).

Lemma 8.2. *Let g be a non-trivial element in $PSL(2, q)$ of order c with f distinct fixed points. Then $c = p$ and $f = 1$, $c \mid \frac{q+1}{n}$ and $f = 0$, or $c \mid \frac{q-1}{n}$ and $f = 2$.*

Lemma 8.3. *Let U be the cyclic group of order c with $c \mid \frac{q\pm1}{n}$. Then, we have*

(a) *if $c \mid \frac{q+1}{n}$, then $N_c = (q+1)/c$,*

(b) *if $c \mid \frac{q-1}{n}$, then $N_1 = 2$ and $N_c = (q-1)/c$.*

Proof. This is an obvious consequence of Lemma 8.2. □

Lemma 8.4. *Let U be the dihedral group of order $2c$ with $c \mid \frac{q\pm1}{n}$. Then*

(i) *for $q \equiv 1 \pmod 4$, we have*

   (a) *if $c \mid \frac{q+1}{2}$, then $N_c = 2$ and $N_{2c} = (q+1-2c)/(2c)$,*

   (b) *if $c \mid \frac{q-1}{2}$, then $N_2 = 1$, $N_c = 2$, and $N_{2c} = (q-1-2c)/(2c)$, unless $c = 2$, in which case $N_2 = 3$ and $N_4 = (q-5)/4$;*

(ii) *for $q \equiv 3 \pmod 4$, we have*

   (a) *if $c \mid \frac{q+1}{2}$, then $N_{2c} = (q+1)/(2c)$,*

   (b) *if $c \mid \frac{q-1}{2}$, then $N_2 = 1$ and $N_{2c} = (q-1)/(2c)$;*

(iii) *for $q \equiv 0 \pmod 2$, we have*

   (a) *if $c \mid q+1$, then $N_c = 1$ and $N_{2c} = (q+1-c)/(2c)$,*

   (b) *if $c \mid q-1$, then $N_2 = 1$, $N_c = 1$, and $N_{2c} = (q-1-c)/(2c)$.*

Proof. First, let $q \equiv 1 \pmod 4$. If $c \mid \frac{q+1}{2}$, then U has a cyclic subgroup of order c, and hence by Lemma 8.3 its orbit-lengths are multiples of c. On the other hand, U has at least c involutions contained in one conjugacy class with two distinct fixed points, and hence we have two orbits of length c and all other orbits are regular. If $c \mid \frac{q-1}{2}$, then U has a cyclic subgroup of order c with two distinct fixed points which are interchanged by an involution, and thus $N_2 \geq 1$. We conclude that $N_2 = 1$, unless $c = 2$, in which case we have exactly three involutions with two distinct fixed points and hence $N_2 = 3$. On the other hand, U has at least c involutions contained in one conjugacy class with two distinct fixed points, and hence we have two orbits of length c if $c > 2$ and all remaining orbits are regular.

For $q \equiv 3 \pmod 4$, we remark that U has at least c involutions contained in one conjugacy class which are fixed point free, and hence we cannot have orbits of length c.

Now, let $q \equiv 0 \pmod 2$. If $c \mid q+1$, then U has a cyclic subgroup of order c, and hence by Lemma 8.3 its orbit-lengths are multiples of c. On the other hand,

U has at least c involutions contained in one conjugacy class with one fixed point, and hence we have one orbit of length c and all remaining orbits are regular. If $c \mid q - 1$, then U has a cyclic subgroup of order c with two distinct fixed points which are interchanged by an involution, thus $N_2 = 1$. On the other hand, U has at least c involutions contained in one conjugacy class with one fixed point, and hence we have one orbit of length c and all remaining orbits are regular. $\qquad\square$

Lemma 8.5. *Let U be the elementary Abelian group of order $\bar{q} \mid q$. Then, we have $N_1 = 1$ and $N_{\bar{q}} = q/\bar{q}$.*

Proof. By the Cauchy-Frobenius Lemma the number of orbits is $(q/\bar{q}) + 1$. As all orbit-lengths are powers of p, we have therefore just one orbit of length 1 and all other orbits are regular. $\qquad\square$

Lemma 8.6. *Let U be a semi-direct product of the elementary Abelian group of order $\bar{q} \mid q$ and the cyclic group of order c with $c \mid \bar{q} - 1$ and $c \mid q - 1$. Then, we have $N_1 = 1$, $N_{\bar{q}} = 1$, and $N_{\bar{q}c} = (q - \bar{q})/(\bar{q}c)$.*

Proof. As U has an elementary Abelian subgroup of order $\bar{q} \mid q$, we can apply Lemma 8.5. Thus, we have one orbit of length 1 and all other orbit-lengths are multiples of $\bar{q}$. However, U has a cyclic subgroup of order c, and thus Lemma 8.3 gives for the orbit-lengths $l \equiv 0$ or 1 (mod c). If $l \equiv 0$ (mod c), then necessarily $l = \bar{q}c$. Otherwise, $l = 1$ or $\bar{q}$. Since an element of order c has two distinct fixed points, the claim follows. $\qquad\square$

Lemma 8.7. *Let U be $PSL(2, \bar{q})$ with $\bar{q}^m = q$, $m \geq 1$. Then, we have $N_{\bar{q}+1} = 1$, $N_{\bar{q}(\bar{q}-1)} = 1$ if m is even, and all other orbits are regular.*

Proof. Due to the fact that all subgroups of the form $PSL(2, \bar{q})$ of $PSL(2, q)$ are conjugate (see, e.g., [44, Chap. 12, p. 279]), U can be identified with the group consisting of all linear fractional maps $GF(\bar{q}) \cup \{\infty\} \longrightarrow GF(\bar{q}) \cup \{\infty\}$, $x \mapsto \frac{ax+b}{cx+d}$ (where $a, b, c, d \in GF(\bar{q})$, $ad - bc$ is a nonzero square and the usual conventions for ∞ holds) with $GF(\bar{q})$ the unique subfield of $GF(q)$ of order $\bar{q}$. As U acts transitively on the points of $GF(\bar{q})$, we have an orbit of length $\bar{q} + 1$. As U has a subgroup of order $\bar{q}(\bar{q} - 1)/n$ which is a semi-direct product of the elementary Abelian group of order $\bar{q} \mid q$ and the cyclic group of order $(\bar{q} - 1)/n$, we deduce from Lemma 8.6 that all other orbit-lengths are multiples of $\bar{q}(\bar{q} - 1)/n$. On the other hand, U has an element of order $(\bar{q} + 1)/n$ which is fixed point free if m is odd, and in this case, all orbit-lengths are multiples of $(\bar{q} + 1)/n$ and hence all other orbits are regular. If m is even, then the element of order $(\bar{q} + 1)/n$ has two distinct fixed points outside the $\bar{q} + 1$ points, and thus we have one orbit of length $\bar{q}(\bar{q} - 1)$ and all remaining orbits are regular. $\qquad\square$

Lemma 8.8. *Let U be $PGL(2, \bar{q})$ with $\bar{q}^m = q$, $m > 1$ even. Then, we have $N_{\bar{q}+1} = 1$, $N_{\bar{q}(\bar{q}-1)} = 1$, and all other orbits are regular.*

Proof. The assertion follows immediately from Lemma 8.7. $\qquad\square$

Lemma 8.9. *Let U be isomorphic to A_4. Then*

(i) *for $q \equiv 1 \pmod 4$, we have*

    (a) *if $3 \mid \frac{q+1}{2}$, then $N_6 = 1$ and $N_{12} = (q-5)/12$,*

    (b) *if $3 \mid \frac{q-1}{2}$, then $N_4 = 2$, $N_6 = 1$, and $N_{12} = (q-13)/12$,*

    (c) *if $3 \mid q$, then $N_4 = 1$, $N_6 = 1$, and $N_{12} = (q-9)/12$;*

(ii) *for $q \equiv 3 \pmod 4$, we have*

    (a) *if $3 \mid \frac{q+1}{2}$, then $N_{12} = (q+1)/12$,*

    (b) *if $3 \mid \frac{q-1}{2}$, then $N_4 = 2$ and $N_{12} = (q-7)/12$,*

    (c) *if $3 \mid q$, then $N_4 = 1$ and $N_{12} = (q-3)/12$;*

(iii) *for $q = 2^e$, $e \equiv 0 \pmod 2$, we have $N_1 = 1$, $N_4 = 1$, and $N_{12} = (q-4)/12$.*

Proof. We have $p > 2$, or $p = 2$ and $e \equiv 0 \pmod 2$. First, let $q \equiv 1 \pmod 4$. As there are in U three involutions contained in one conjugacy class with two distinct fixed points, we have always one orbit of length 6 in subcases (a) and (b). On the other hand, there are in U four subgroups of order 3 contained in one conjugacy class with two distinct fixed points if $3 \mid \frac{q-1}{2}$, and none if $3 \mid \frac{q+1}{2}$. This implies for (a) that all remaining orbits are regular, and for (b) that U has exactly two orbits of equal length on the set of these fixed points and all other orbits are regular. If $3 \mid q$, then we have more precisely $q = 3^e$ with $e \equiv 0 \pmod 2$, and since $PSL(2,3) \cong A_4$, the claim follows by applying Lemma 8.7.

For $q \equiv 3 \pmod 4$, we remark that the three involutions contained in one conjugacy class are fixed point free, and hence we cannot have orbits of length 6 in (a) and (b). If $3 \mid q$, then we have more precisely $q = 3^e$ with $e \equiv 1 \pmod 2$, and the claim follows again by Lemma 8.7.

Now, let $q = 2^e$ with $e \equiv 0 \pmod 2$. Since the set of fixed points of some subgroup is left invariant by its normalizer, clearly the normalizer $\mathcal{N}_U(P)$ of a Sylow 2-subgroup P in U has then exactly one fixed point. But as in U there is only one Sylow 2-subgroup, we have clearly $\mathcal{N}_U(P) = U$, and hence U has one orbit of length 1. On the other hand, since always $3 \mid q - 1$ in this case, U has an element of order 3 with two distinct fixed points which implies the existence of one orbit of length 4, and all remaining orbits are regular. $\qquad\square$

Lemma 8.10. *Let U be isomorphic to S_4. Then*

(i) *for $q \equiv 1 \pmod 8$, we have*

    (a) *if $3 \mid \frac{q+1}{2}$, then $N_6 = 1$, $N_{12} = 1$, and $N_{24} = (q-17)/24$,*

    (b) *if $3 \mid \frac{q-1}{2}$, then $N_6 = 1$, $N_8 = 1$, $N_{12} = 1$, and $N_{24} = (q-25)/24$,*

(c) *if $3 \mid q$, then $N_4 = 1$, $N_6 = 1$, and $N_{24} = (q-9)/24$;*

(ii) *for $q \equiv -1 \pmod 8$, we have*

(a) *if $3 \mid \frac{q+1}{2}$, then $N_{24} = (q+1)/24$,*

(b) *if $3 \mid \frac{q-1}{2}$, then $N_8 = 1$ and $N_{24} = (q-7)/24$.*

Proof. We have $q \equiv \pm 1 \pmod 8$. As U has a subgroup isomorphic to A_4, Lemma 8.9 gives orbits of length $4, 6, 8, 12, 24$. First, let $q \equiv 1 \pmod 8$. As there are in U three involutions contained in one conjugacy class with two distinct fixed points, we have always one orbit of length 6 in subcases (a) and (b). On the other hand, we have in U four subgroups of order 3 contained in one conjugacy class with two distinct fixed points if $3 \mid \frac{q-1}{2}$, and none if $3 \mid \frac{q+1}{2}$. Thus, for (b) we conclude that U necessarily has one orbit of length 8 on the set of these fixed points. Furthermore, if $N_{12} = 0$, then $N_{24} = (q-13)/24$ which is not integer. Hence, $N_{12} = 1$ and $N_{24} = (q-25)/24$. For (a) we deduce again if $N_{12} = 0$, then $N_{24} = (q-5)/24$ which is not integer. Thus, $N_{12} = 1$ and $N_{24} = (q-17)/24$. For $3 \mid q$, the assertion follows obviously from Lemma 8.9 (ii)(c).

Now, let $q \equiv -1 \pmod 8$. Then, clearly $3 \nmid q$. We remark that the three involutions contained in one conjugacy class are fixed point free, and hence we cannot have orbits of length 6 in (a) and (b). Furthermore, we cannot have orbits of length 12 since otherwise we would have one-point stabilizers of order 2. $\square$

Lemma 8.11. *Let U be isomorphic to A_5. Then*

(i) *for $q \equiv 1 \pmod 4$, we have*

(a) *if $q = 5^e$, $e \equiv 1 \pmod 2$, then $N_6 = 1$ and $N_{60} = (q-5)/60$,*

(b) *if $q = 5^e$, $e \equiv 0 \pmod 2$, then $N_6 = 1$, $N_{20} = 1$, and $N_{60} = (q-25)/60$,*

(c) *if $15 \mid \frac{q+1}{2}$, then $N_{30} = 1$ and $N_{60} = (q-29)/60$,*

(d) *if $3 \mid \frac{q+1}{2}$ and $5 \mid \frac{q-1}{2}$, then $N_{12} = 1$, $N_{30} = 1$, and $N_{60} = (q-41)/60$,*

(e) *if $3 \mid \frac{q-1}{2}$ and $5 \mid \frac{q+1}{2}$, then $N_{20} = 1$, $N_{30} = 1$, and $N_{60} = (q-49)/60$,*

(f) *if $15 \mid \frac{q-1}{2}$, then $N_{12} = 1$, $N_{20} = 1$, $N_{30} = 1$, and $N_{60} = (q-61)/60$,*

(g) *if $3 \mid q$ and $5 \mid \frac{q+1}{2}$, then $N_{10} = 1$ and $N_{60} = (q-9)/60$,*

(h) *if $3 \mid q$ and $5 \mid \frac{q-1}{2}$, then $N_{10} = 1$, $N_{12} = 1$, and $N_{60} = (q-21)/60$;*

(ii) *for $q \equiv 3 \pmod 4$, we have*

(a) *if $15 \mid \frac{q+1}{2}$, then $N_{60} = (q+1)/60$,*

(b) *if $3 \mid \frac{q+1}{2}$ and $5 \mid \frac{q-1}{2}$, then $N_{12} = 1$ and $N_{60} = (q-11)/60$,*

(c) *if $3 \mid \frac{q-1}{2}$ and $5 \mid \frac{q+1}{2}$, then $N_{20} = 1$ and $N_{60} = (q-19)/60$,*

(d) *if $15 \mid \frac{q-1}{2}$, then $N_{12} = 1$, $N_{20} = 1$, and $N_{60} = (q-31)/60$.*

Proof. We have $p = 5$ or $q \equiv \pm 1 \pmod{10}$. We note that U has a subgroup isomorphic to A_4. Let $q \equiv 1 \pmod 4$. For $p = 5$, we have $PSL(2,5) \cong A_5$, and thus assertions (a) and (b) follow from Lemma 8.7. For the remaining subcases we distinguish the cases $3 \mid \frac{q \pm 1}{2}$ or $3 \mid q$, and $5 \mid \frac{q \pm 1}{2}$.

ad (c): By Lemma 8.9 (ii)(a), all orbit-lengths are multiples of 6 respectively 12. On the other hand, U has a fixed point free element of order 5, which means that all orbit-lengths are multiples of 5. Thus, we have $N_{30} = 1$ and all other orbits are regular.

ad (d): Again, by Lemma 8.9 (ii)(a), all orbit-lengths are multiples of 6 respectively 12. Furthermore, we cannot have an orbit of length 6 since otherwise we would have a one-point stabilizer of order 10, which is not possible for $p \neq 5$ as in A_5 all subgroups of order 10 are isomorphic to dihedral groups. On the other hand, U has an element of order 5 with two distinct fixed points which implies the existence of one orbit of size 12. Therefore, we have $N_{12} = 1$, $N_{30} = 1$, and all remaining orbits are regular.

ad (e): We deduce from Lemma 8.9 (ii)(b) that all orbit-lengths are multiples of 4, 6 respectively 12. On the other hand, U has a fixed point free element of order 5, which means that all orbit-lengths are multiples of 5. Hence, we conclude that $N_{20} = 1$, $N_{30} = 1$ and all other orbits are regular.

ad (f): By Lemma 8.9 (ii)(b) again, all orbit-lengths are multiples of 4, 6 respectively 12. We may conclude as in (d) that $N_6 = 0$. Furthermore, we cannot have an orbit of length 4 since otherwise we would have a one-point stabilizer of order 15, which is impossible as the non-Abelian simple group A_5 has proper subgroups only of index at least 5. On the other hand, U has an element of order 5 with two distinct fixed points which implies the existence of one orbit of size 12. Therefore, we have $N_{12} = 1$, $N_{20} = 1$, $N_{30} = 1$, and all remaining orbits are regular.

Let $3 \mid q$. Since the set of fixed points of some subgroup is left invariant by its normalizer, clearly the normalizer $\mathcal{N}_U(P)$ of a Sylow 3-subgroup P in U has then exactly one fixed point. As we have 10 Sylow 3-subgroups in U contained in one conjugacy class, we conclude that $|\mathcal{N}_U(P)| = 6$. Since $\mathcal{N}_U(P)$ is a maximal subgroup in U, it follows therefore that we have one orbit of length 10. If $5 \mid \frac{q+1}{2}$, then U has a fixed point free element of order 5, and hence it follows that all other orbits are regular. If $5 \mid \frac{q-1}{2}$, then U has an element of order 5 with two distinct fixed points which implies the existence of one orbit of size 12 since $N_6 = 0$ as in (d), and all remaining orbits are regular.

For $q \equiv 3 \pmod 4$, clearly $p = 5$ is not possible, and hence it follows that $3 \nmid q$ and $5 \mid \frac{q \pm 1}{2}$. Since a subgroup of U which is isomorphic to A_4 cannot have orbits of length 6 due to Lemma 8.9 (ii), we may proceed, mutatis mutandis, as in subcases (c)–(f) above. $\qquad \square$

We shall now turn to the examination of those cases where $G \leq \mathrm{Aut}(\mathcal{D})$ is of almost simple type.

Case (1): $N = A_v$, $v \geq 5$.

We may assume that $v \geq 6$. But then A_v, and hence also G, is 4-transitive and does not act on any non-trivial Steiner 4-design $\mathcal{D}$ in view of Theorem 5.2.

Case (2): $N = PSL(d, q)$, $d \geq 2$, $v = \frac{q^d - 1}{q - 1}$, where $(d, q) \neq (2, 2), (2, 3)$.

We distinguish two subcases:

Case (2a): $N = PSL(2, q)$, $v = q + 1$, $q = p^e > 3$.

Here $\operatorname{Aut}(N) = P\Gamma L(2, q)$, and $|G| = (q + 1)q\frac{(q-1)}{n}a$ with $n = (2, q - 1)$ and $a \mid ne$. We may assume that $q \geq 5$. We will show that $G \leq \operatorname{Aut}(\mathcal{D})$ cannot act flag-transitively on any non-trivial Steiner 4-design $\mathcal{D}$.

First, assume that $N = G$. Then, by Remark 4.15, we obtain

$$(q - 2)|PSL(2, q)_{0B}| \cdot n = (k - 1)(k - 2)(k - 3) \tag{8.1}$$

which is equivalent to

$$(q - 2)|PSL(2, q)_{0B}| \cdot n + 6 = k(k^2 - 6k + 11). \tag{8.2}$$

Thus, we have in particular

$$k \mid (q - 2)|PSL(2, q)_{0B}| \cdot n + 6. \tag{8.3}$$

Since $PSL(2, q)_B$ acts transitively on the points of B, we have

$$k = \left|0^{PSL(2,q)_B}\right| = [PSL(2, q)_B : PSL(2, q)_{0B}]. \tag{8.4}$$

Let us first consider the case when $|PSL(2, q)_{0B}| = 1$. If q is even, then $k \mid q + 4$ by property (8.3). On the other hand, using equation (8.4), we have $k = |PSL(2, q)_B| \mid |PSL(2, q)| = q^3 - q$. As

$$(q^3 - q, q + 4) = (60, q + 4) = 4 \cdot (15, 2^{e-2} + 1) = \begin{cases} 4, & \text{if } e \text{ is even and } 4 \nmid e, \\ 4 \cdot 3, & \text{if } e \text{ is odd}, \\ 4 \cdot 5, & \text{if } 4 \mid e, \end{cases}$$

the possible values for k can immediately be ruled out by hand using equation (8.1). If q is odd, we have $k = |PSL(2, q)_B| \mid 2(q + 1)$ due to property (8.3) and equation (8.4). Examining the list of subgroups of $PSL(2, q)$ (cf. [44, Chap. 12, p. 285f.] or [73, Chap. II, Thm. 8.27]), we have to consider the following possibilities:

(i) $PSL(2, q)_B$ is conjugate to a cyclic subgroup of order c with $c \mid \frac{q+1}{2}$ of $PSL(2, q)$, and $k = c$.

(ii) $PSL(2, q)_B$ is conjugate to a dihedral subgroup of order $2c$ with $c \mid \frac{q+1}{2}$ of $PSL(2, q)$, and $k = 2c$.

(iii) $PSL(2, q)_B$ is conjugate to A_4, and $k = 12$.

(iv) $PSL(2,q)_B$ is conjugate to S_4, and $k = 24$.

(v) $PSL(2,q)_B$ is conjugate to A_5, and $k = 60$.

ad (i): By equation (8.1), we have

$$c \mid \frac{q+1}{2} = \frac{(c-1)(c-2)(c-3)+6}{4} = \frac{c(c^2-6c+11)}{4}.$$

Since 4 does not divide $c^2 - 6c + 11$, this is impossible.

ad (ii): Using equation (8.1), we obtain

$$c \mid \frac{q+1}{2} = \frac{(2c-1)(c-1)(2c-3)+3}{2},$$

which is not possible as $(2c-1)(c-1)(2c-3) + 3 = 4c^3 - 12c^2 + 11c \equiv c \pmod{2c}$.

ad (iii)–(v): For each given value of k, equation (8.1) gives in each subcase that q is not a prime power.

We consider now the case when $|PSL(2,q)_{0B}| = 2$. If q is even, then we have $k = \frac{|PSL(2,q)_B|}{2} \mid 2(q+1)$ due to property (8.3) and equation (8.4). Considering the list of subgroups of $PSL(2,q)$, we have the following possibilities:

(i) $PSL(2,q)_B$ is conjugate to a cyclic subgroup of order c with $c \mid q+1$ of $PSL(2,q)$, and $k = \frac{c}{2}$.

(ii) $PSL(2,q)_B$ is conjugate to a dihedral subgroup of order $2c$ with $c \mid q+1$ of $PSL(2,q)$, and $k = c$.

(iii) $PSL(2,q)_B$ is conjugate to $PSL(2,\bar q)$ with $\bar q \mid 4$, and $k = 30$.

(iv) $PSL(2,q)_B$ is conjugate to A_4, and $k = 6$.

ad (i), (iii), (iv): In view of Lemmas 8.3, 8.8, respectively 8.9 (iii), clearly k cannot take the given values.

ad (ii): Considering equation (8.1), for $k = c > 4$ clearly the right-hand side of the equation is divisible by 8, but not the left-hand side.

If q is odd, then $k \mid 2(2q - 1)$ by property (8.3). On the other hand, using equation (8.4) gives $k = \frac{|PSL(2,q)_B|}{2} \mid \frac{|PSL(2,q)|}{2} = \frac{q^3-q}{4}$. Since $\left(\frac{q^3-q}{4}, 2(2q-1)\right) = 2 \cdot \left(\frac{q^3-q}{8}, 2q - 1\right) = 2 \cdot (3, q+1)$ only $k = 6$ can occur, and equation (8.1) gives then $q = 17$. However, it is known that there does not exist any 4-(18, 6, 1) design (cf. [126, Thm. 6]).

Finally, let us consider the case when $|PSL(2,q)_{0B}| > 2$. Examining the list of subgroups of $PSL(2,q)$ with their orbits on the projective line (Lemmas 8.3-8.11), we have to consider the following subcases:

(i) $PSL(2,q)_B$ is conjugate to S_4, and $k = 6$ or 8.

(ii) $PSL(2,q)_B$ is conjugate to A_5, and $k = 6, 10, 12$ or 20.

(iii) $PSL(2,q)_B$ is conjugate to a semi-direct product of an elementary Abelian subgroup of order $\bar{q} \mid q$ with a cyclic subgroup of order c of $PSL(2,q)$ with $c \mid \bar{q} - 1$ and $c \mid q - 1$, and $k = \bar{q}$.

(iv) $PSL(2,q)_B$ is conjugate to $PSL(2,\bar{q})$ with $\bar{q}^m = q$, $m \geq 1$, and $k = \bar{q} + 1$ or $\bar{q}(\bar{q} - 1)$ if m is even.

(v) $PSL(2,q)_B$ is conjugate to $PGL(2,\bar{q})$ with $\bar{q}^m = q$, $m > 1$ even, and $k = \bar{q} + 1$ or $\bar{q}(\bar{q} - 1)$.

ad (i): We may assume that q is odd. Applying equations (8.1) and (8.4) implies for $k = 6$ that q is not a prime power, and for $k = 8$ that $q = 37$, in which case $q \equiv \pm 1 \pmod 8$ (cf. Lemma 8.10) does not hold.

ad (ii): Again, we may assume that q is odd and consider equations (8.1) and (8.4) for the given values of k. We obtain for $k = 6$ that $q = 5$, which is clearly impossible due to Corollary 1.17, for $k = 10$ that q is not a prime power, and for $k = 20$ that $q = 971$, in which case Lemma 1.14 (c) gives a contradiction. If $k = 12$, then we get $q = 101$. Since $|PSL(2,q)_{0B}| = 5$ by equation (8.4) and $5 \mid \frac{q-1}{2}$, $PSL(2,q)_{0B}$ has two distinct fixed points. If one fixed point lies outside B, then clearly $q \equiv 1 \pmod 5$ and hence $k = 12$ is not possible. Thus, we may assume that both fixed points are incident with B. But then, as every non-identity element of $PSL(2,q)$ fixes at most two distinct points, $PSL(2,q)_{0B}$ must fix some 2-subset by the definition of Steiner 4-designs, and hence contains an involution, a contradiction.

ad (iii): We have $((q-2)|PSL(2,q)_{0B}| \cdot n + 6, q) = (2 \cdot |PSL(2,q)_{0B}| \cdot n - 6, q)$, and property (8.3) gives in particular

$$k \mid 2 \cdot |PSL(2,q)_{0B}| \cdot n - 6. \tag{8.5}$$

On the other hand, it follows from equation (8.4) that $|PSL(2,q)_{0B}| \mid k - 1$. Therefore, we have in particular

$$\frac{k-1}{2n} < \frac{k+6}{2n} \leq |PSL(2,q)_{0B}| \mid k - 1. \tag{8.6}$$

If q is even, then we deduce that $|PSL(2,q)_{0B}| = k - 1$. Property (8.5) yields then $k \mid 2k - 8$, and, as clearly $(2k - 8, k) = (8, k)$, only $k = 8$ is possible. Thus, we have $q = 32$ in view of equation (8.1), which is impossible by Lemma 1.14 (c). If q is odd, then by property (8.6), we have to consider the possibilities when $|PSL(2,q)_{0B}| = \frac{k-1}{\bar{n}}$ with $\bar{n} = 1, 2, 3$. If $|PSL(2,q)_{0B}| = k - 1$, then we obtain $k \mid 4k - 10$ by property (8.3). Clearly, $(4k - 10, k) = (10, k)$, but as $k \mid q$, only $k = 5$ is possible. Then, equation (8.1) gives $q = 5$, which leads to a contradiction in view of Corollary 1.17. For $|PSL(2,q)_{0B}| = \frac{k-1}{2}$, we have $k \mid 2k - 8$ and thus $k = 8$ as above, which is impossible as $k \nmid q$. If $|PSL(2,q)_{0B}| = \frac{k-1}{3}$, then property (8.3) gives $k \mid \frac{4k-22}{3}$. Since $(4k - 22, 3k) = (22, 3k)$ is not divisible by 3, this is not possible.

ad (iv) and (v): In subcase (iv), we have $|PSL(2,q)_{0B}| = \frac{\bar{q}(\bar{q}-1)}{n}$ if $k = \bar{q}+1$. Thus, equation (8.1) implies for $k = \bar{q}+1$ that $q = \bar{q}$ must hold, which is impossible due to Corollary 1.17. For $m > 1$ even and $k = \bar{q}(\bar{q}-1)$, it follows that $|PSL(2,q)_{0B}| = \frac{\bar{q}+1}{n}$. Hence, by property (8.3), we conclude that

$$\bar{q}(\bar{q}-1) \mid (q-2)(\bar{q}+1) + 6 = \bar{q}^{m+1} + \bar{q}^m - 2\bar{q} + 4.$$

Since $(\bar{q}^{m+1} + \bar{q}^m - 2\bar{q} + 4, \bar{q}) = (4, \bar{q})$ and $k > 4$, only the case when $\bar{q} = 4$ has to be considered. Thus, $k = 12$ and applying equation (8.2) immediately gives a contradiction. In subcase (v), clearly n does not appear in either equations (8.1) and (8.2) or in property (8.3), and thus we may argue mutatis mutandis as in subcase (iv).

Now, let us assume that $N < G \leq \mathrm{Aut}(N)$. We recall that $q = p^e > 3$, and will distinguish in the following the cases $p > 2$ and $p = 2$.

First, let $p > 2$. We define

$$G^* = G \cap (PSL(2,q) \rtimes \langle \tau_\alpha \rangle)$$

with $\tau_\alpha \in \mathrm{Sym}(GF(p^e) \cup \{\infty\}) \cong S_v$ of order e induced by the Frobenius automorphism $\alpha : GF(p^e) \longrightarrow GF(p^e)$, $x \mapsto x^p$. Then, by Dedekind's law, we can write

$$G^* = PSL(2,q) \rtimes (G^* \cap \langle \tau_\alpha \rangle). \tag{8.7}$$

Defining $P\Sigma L(2,q) = PSL(2,q) \rtimes \langle \tau_\alpha \rangle$, it can easily be calculated that $P\Sigma L(2,q)_{0,1,\infty} = \langle \tau_\alpha \rangle$, and $\langle \tau_\alpha \rangle$ has precisely $p + 1$ distinct fixed points (cf., e.g., [43, Chap. 6.4, Lemma 2]). As $p > 2$, we conclude therefore that $G^* \cap \langle \tau_\alpha \rangle \leq G^*_{0B}$ for some appropriate, unique block $B \in \mathcal{B}$ by the definition of Steiner 4-designs. Furthermore, clearly $PSL(2,q) \cap (G^* \cap \langle \tau_\alpha \rangle) = 1$. Hence, we have

$$\begin{aligned}
\left|(0,B)^{G^*}\right| &= [G^* : G^*_{0B}] \\
&= [PSL(2,q) \rtimes (G^* \cap \langle \tau_\alpha \rangle) : PSL(2,q)_{0B} \rtimes (G^* \cap \langle \tau_\alpha \rangle)] \\
&= [PSL(2,q) : PSL(2,q)_{0B}] \\
&= \left|(0,B)^{PSL(2,q)}\right|.
\end{aligned} \tag{8.8}$$

Thus, if we assume that $G^* \leq \mathrm{Aut}(\mathcal{D})$ acts already flag-transitively on $\mathcal{D}$, then we obtain $\left|(0,B)^{G^*}\right| = \left|(0,B)^{PSL(2,q)}\right| = bk$ in view of Remark 4.15. Hence, $PSL(2,q)$ must also act flag-transitively on $\mathcal{D}$, and we may proceed as in the case when $N = G$. Therefore, let us assume that $G^* \leq \mathrm{Aut}(\mathcal{D})$ does not act flag-transitively on $\mathcal{D}$. Then, $[G : G^*] = 2$ and G^* has exactly two orbits of equal length on the set of flags. Thus, by equation (8.8), we obtain for the orbit containing the flag $(0,B)$ that $\left|(0,B)^{G^*}\right| = \left|(0,B)^{PSL(2,q)}\right| = \frac{bk}{2}$. As $PSL(2,q)$ is normal in G, we have under $PSL(2,q)$ also precisely one further orbit of equal length on the set of flags. Then, proceeding similarly to the case $N = G$ for each orbit on the set of flags, we obtain (representative for the orbit containing the flag $(0, B)$) that

$$\frac{(q-2)\,|PSL(2,q)_{0B}|\cdot n}{2} = (k-1)(k-2)(k-3) \tag{8.9}$$

which is equivalent to

$$\frac{(q-2)\,|PSL(2,q)_{0B}|\cdot n}{2} + 6 = k(k^2 - 6k + 11). \tag{8.10}$$

Hence, we have in particular

$$k \left| \frac{(q-2)\,|PSL(2,q)_{0B}|\cdot n}{2} + 6. \right. \tag{8.11}$$

Since $PSL(2,q)_B$ can have one or two orbits of equal length on the points of B, we have

$$k \text{ or } \frac{k}{2} = \left| 0^{PSL(2,q)_B} \right| = [PSL(2,q)_B : PSL(2,q)_{0B}]. \tag{8.12}$$

Let us recall that here q is always odd. First considering the case when $|PSL(2,q)_{0B}| = 1$ gives immediately a contradiction to equation (8.9). Let us now observe the case when $|PSL(2,q)_{0B}| = 2$. We have $k \mid 2(q+1)$ in view of property (8.11), and $k = |PSL(2,q)_B|$ or $\frac{|PSL(2,q)_B|}{2}$ by equation (8.12). For $k = |PSL(2,q)_B|$, clearly equation (8.9) with $|PSL(2,q)_{0B}| = 2$ is equivalent to equation (8.1) with $|PSL(2,q)_{0B}| = 1$, and thus we can argue exactly as in case $N = G$ for $|PSL(2,q)_{0B}| = 1$ and q odd. For $k = \frac{|PSL(2,q)_B|}{2}$, we have to consider the following subgroups of $PSL(2,q)$:

(i) $PSL(2,q)_B$ is conjugate to a cyclic subgroup of order c with $c \mid \frac{q+1}{2}$ of $PSL(2,q)$, and $k = \frac{c}{2}$.

(ii) $PSL(2,q)_B$ is conjugate to a dihedral subgroup of order $2c$ with $c \mid \frac{q+1}{2}$ of $PSL(2,q)$, and $k = c$.

(iii) $PSL(2,q)_B$ is conjugate to A_4, and $k = 6$.

(iv) $PSL(2,q)_B$ is conjugate to S_4, and $k = 12$.

(v) $PSL(2,q)_B$ is conjugate to A_5, and $k = 30$.

ad (i): Obviously, k cannot take the given value due to Lemma 8.3.
ad (ii): It follows from equation (8.9) that

$$c \left| \frac{q+1}{2} = \frac{(c-1)(c-2)(c-3)+6}{4} = \frac{c(c^2 - 6c + 11)}{4}. \right.$$

As 4 does not divide $c^2 - 6c + 11$, this is impossible.
ad (iii)–(v): In view of equation (8.9), we obtain in subcase (iii) that $q = 32$, which is not possible, and in each of the other subcases that q is not a prime power.

We consider finally the case when $|PSL(2,q)_{0B}| > 2$. Combining equations (8.9) and (8.12), we obtain

$$\frac{(q-2)\,|PSL(2,q)_B|\cdot n}{2} = k(k-1)(k-2)(k-3) \tag{8.13}$$

$$\text{with } k = \left|0^{PSL(2,q)_B}\right| = \frac{|PSL(2,q)_B|}{|PSL(2,q)_{0B}|}, \text{ or}$$

$$(q-2)\,|PSL(2,q)_B|\cdot n = k(k-1)(k-2)(k-3) \tag{8.14}$$

$$\text{with } k = 2\cdot\left|0^{PSL(2,q)_B}\right| = 2\cdot\frac{|PSL(2,q)_B|}{|PSL(2,q)_{0B}|}.$$

In view of the subgroups of $PSL(2,q)$ with their orbits on the projective line (Lemmas 8.3-8.11), we have the following possibilities:

(i) $PSL(2,q)_B$ is conjugate to A_4, and $k = 2\cdot\left|0^{PSL(2,q)_B}\right| = 8$.

(ii) $PSL(2,q)_B$ is conjugate to S_4, and $k = \left|0^{PSL(2,q)_B}\right| = 6$ or 8, respectively $k = 2\cdot\left|0^{PSL(2,q)_B}\right| = 8, 12$ or 16.

(iii) $PSL(2,q)_B$ is conjugate to A_5, and $k = \left|0^{PSL(2,q)_B}\right| = 6, 10, 12$ or 20, respectively $k = 2\cdot\left|0^{PSL(2,q)_B}\right| = 12, 20, 24$ or 40.

(iv) $PSL(2,q)_B$ is conjugate to a semi-direct product of an elementary Abelian subgroup of order $\bar{q}\mid q$ with a cyclic subgroup of order c of $PSL(2,q)$ with $c\mid\bar{q}-1$ and $c\mid q-1$, and $k = \left|0^{PSL(2,q)_B}\right| = \bar{q}$, respectively $k = 2\cdot\left|0^{PSL(2,q)_B}\right| = 2\bar{q}$.

(v) $PSL(2,q)_B$ is conjugate to $PSL(2,\bar{q})$ with $\bar{q}^m = q$, $m \geq 1$, and $k = \left|0^{PSL(2,q)_B}\right| = \bar{q}+1$, or $\bar{q}(\bar{q}-1)$ if m is even, respectively $k = 2\cdot\left|0^{PSL(2,q)_B}\right| = 2(\bar{q}+1)$, or $2\bar{q}(\bar{q}-1)$ if m is even.

(vi) $PSL(2,q)_B$ is conjugate to $PGL(2,\bar{q})$ with $\bar{q}^m = q$, $m > 1$ even, and $k = \left|0^{PSL(2,q)_B}\right| = \bar{q}+1$ or $\bar{q}(\bar{q}-1)$, respectively $k = 2\cdot\left|0^{PSL(2,q)_B}\right| = 2(\bar{q}+1)$ or $2\bar{q}(\bar{q}-1)$.

ad (i): By equation (8.14), we obtain that q is not a prime power.

ad (ii): First, applying equation (8.13) implies for $k = 6$ that $q = 17$, which can be excluded since there does not exist any 4-(18,6,1) design as already mentioned, and for $k = 8$ that q is not a prime power. Using equation (8.14) gives for $k = 8$ that $q = 37$, in which case $q \equiv \pm 1 \pmod 8$ (cf. Lemma 8.10) does not hold, and for $k = 12$ and 16 that q is not a prime power in each case.

ad(iii): Observing first equation (8.13) gives for each given value of k that q would be even. Now, applying equation (8.14) gives for $k = 12$ the prime $q = 101$, which is impossible as according to Lemma 8.11 we only have orbits of length 6 when $p = 5$, and for $k = 20$ that $q = 971$, in which case Lemma 1.14 (c) gives a contradiction. For $k = 24$ and 40, we obtain in each case that q is not a prime power.

ad (iv): Let $k = \bar{q}$. As $\left(\frac{(q-2)|PSL(2,q)_{0B}|\cdot n}{2} + 6, q\right) = \left(|PSL(2,q)_{0B}|\cdot n - 6, q\right)$, property (8.11) implies that

$$k \mid |PSL(2,q)_{0B}| \cdot n - 6. \tag{8.15}$$

On the other hand, as $k = |0^{PSL(2,q)_B}| = [PSL(2,q)_B : PSL(2,q)_{0B}]$ in this case, it follows that $|PSL(2,q)_{0B}| = c \mid k - 1$. Thus, in particular

$$\frac{k-1}{n} < \frac{k+6}{n} \le |PSL(2,q)_{0B}| \mid k - 1,$$

and hence $|PSL(2,q)_{0B}| = k - 1$ as q is odd. But, property (8.15) gives $k \mid 2k - 8$, and as clearly $(2k - 8, k) = (8, k)$, it would follow that $k = 8$, which is impossible for q odd. For $k = 2\bar{q}$, it follows from equation (8.14) that

$$(q - 2)n = 4 \cdot \frac{(\bar{q} - 1)}{c}(2\bar{q} - 1)(2\bar{q} - 3),$$

which gives a contradiction as clearly the left-hand side of the equation is not divisible by 4.

ad (v) and (vi): We first consider subcase (v). For $k = \bar{q} + 1$, it follows from equation (8.13) that $q = 2(\bar{q} - 1)$, which is obviously impossible for $\bar{q} > 2$. If $m > 1$ even and $k = \bar{q}(\bar{q} - 1)$, then we have

$$(q - 2)(\bar{q} + 1) = 2(\bar{q}^2 - \bar{q} - 1)(\bar{q}^2 - \bar{q} - 2)(\bar{q}^2 - \bar{q} - 3)$$

in view of equation (8.13). As clearly $(\bar{q}^2 - \bar{q} - 1, \bar{q} + 1) = 1$, it follows that $\bar{q}^2 - \bar{q} - 1 \mid q - 2$ must hold. But, polynomial division with remainder gives

$$q - 2 = \left(\sum_{i=1}^{m-1} n_i \frac{q}{\bar{q}^{i+1}}\right)\left(\bar{q}^2 - \bar{q} - 1\right) + n_m\bar{q} + n_{m-1} - 2, \tag{8.16}$$

where n_i denotes the i-th Fibonacci number recursively defined via

$$n_1 = n_2 = 1, \quad n_i = n_{i-1} + n_{i-2} \ (i \ge 3).$$

Hence, as clearly $n_m\bar{q} + n_{m-1} - 2 > 0$ for $m > 1$, we obtain a contradiction. If $k = 2(\bar{q} + 1)$, then applying equation (8.14) gives

$$(q - 2)(\bar{q} - 1) = 4(2\bar{q} + 1)(2\bar{q} - 1). \tag{8.17}$$

As clearly $(2\bar{q} - 1, \bar{q} - 1) = 1$, we deduce that $2\bar{q} - 1 \mid q - 2$ must hold. Since polynomial division with remainder gives

$$q - 2 = \left(\sum_{i=1}^{\overline{m}} \frac{q}{(2\bar{q})^i}\right)\left(2\bar{q} - 1\right) + \frac{q}{(2\bar{q})^{\overline{m}}} - 2$$

for a suitable $\overline{m} \in \mathbb{N}$ (such that

$$\deg\left(\frac{q}{(2\overline{q})^{\overline{m}}} - 2\right) < \deg\left(2\overline{q} - 1\right)$$

as is well-known), it follows therefore that $\overline{q}$, and hence also q, is necessarily a power of 2, a contradiction. For $m > 1$ even and $k = 2\overline{q}(\overline{q} - 1)$, equation (8.14) gives

$$(q - 2)(\overline{q} + 1) = 2(2\overline{q}^2 - 2\overline{q} - 1)2(\overline{q}^2 - \overline{q} - 1)(2\overline{q}^2 - 2\overline{q} - 3).$$

Again, as obviously $(\overline{q}^2 - \overline{q} - 1, \overline{q} + 1) = 1$, we deduce that $\overline{q}^2 - \overline{q} - 1 \mid q - 2$ must hold, and we may proceed exactly as above for $k = \overline{q}(\overline{q} - 1)$. In subcase (vi), clearly n does not appear in equations (8.13) and (8.14), and we may argue mutatis mutandis as in subcase (v).

Now, let $p = 2$. Then, clearly $N = PSL(2, q) = PGL(2, q)$, and we have $\mathrm{Aut}(N) = P\Sigma L(2, q)$. If we assume that $\langle \tau_\alpha \rangle \leq P\Sigma L(2, q)_{0B}$ for some appropriate, unique block $B \in \mathcal{B}$, then, using the terminology of (8.7), we have $G^* = G = P\Sigma L(2, q)$ and as clearly $PSL(2, q) \cap \langle \tau_\alpha \rangle = 1$, we can apply equation (8.8). Thus, $PSL(2, q)$ must also be flag-transitive, which has already been considered. Therefore, we may assume that $\langle \tau_\alpha \rangle \nleq P\Sigma L(2, q)_{0B}$. Let s be a prime divisor of $e = |\langle \tau_\alpha \rangle|$. As the normal subgroup $H := (P\Sigma L(2, q)_{0,1,\infty})^s \leq \langle \tau_\alpha \rangle$ of index s fixes at least four distinct points, we have $G \cap H \leq G_{0B}$ for some appropriate, unique block $B \in \mathcal{B}$ by the definition of Steiner 4-designs. It can then be deduced that $e = s^u$ for some $u \in \mathbb{N}$, since if we assume for $G = P\Sigma L(2, q)$ that there exists a further prime divisor $\overline{s}$ of e with $\overline{s} \neq s$, then $\overline{H} := (P\Sigma L(2, q)_{0,1,\infty})^{\overline{s}} \leq \langle \tau_\alpha \rangle$ and H are both subgroups of $P\Sigma L(2, q)_{0B}$ by the flag-transitivity of $P\Sigma L(2, q)$, and hence $\langle \tau_\alpha \rangle \leq P\Sigma L(2, q)_{0B}$, a contradiction. Furthermore, as $\langle \tau_\alpha \rangle \nleq P\Sigma L(2, q)_{0B}$, we may, by applying Dedekind's law, assume that

$$G_{0B} = PSL(2, q)_{0B} \rtimes (G \cap H).$$

Thus, by Remark 4.15, we obtain

$$(q - 2)|PSL(2, q)_{0B}||G \cap H| = (k - 1)(k - 2)(k - 3)|G \cap \langle \tau_\alpha \rangle|.$$

Using that $k = |0^{G_B}| = [G_B : G_{0B}]$, we have more precisely

(A) if $G = PSL(2, q) \rtimes (G \cap H)$:

$$(q - 2)|PSL(2, q)_{0B}| = (k - 1)(k - 2)(k - 3)$$

$$\text{with } |PSL(2, q)_{0B}| = \frac{|PSL(2, q)_B|}{k}, \quad \text{or}$$

(B) if $G = P\Sigma L(2, q)$:

$$(q - 2)|PSL(2, q)_{0B}| = (k - 1)(k - 2)(k - 3)s$$

$$\text{with } |PSL(2, q)_{0B}| = \frac{|PSL(2, q)_B|}{k} \cdot \begin{cases} s, & \text{if } G_B = PSL(2, q)_B \rtimes \langle \tau_\alpha \rangle, \\ 1, & \text{if } G_B = PSL(2, q)_B \rtimes H. \end{cases}$$

As far as condition (A) is concerned, we may argue exactly as in the earlier case $N = G$ for q even. Thus, only condition (B) has to be examined, and we will also show that here $G \leq \mathrm{Aut}(\mathcal{D})$ cannot act flag-transitively on any non-trivial Steiner 4-design $\mathcal{D}$. Clearly, for each $B \in \mathcal{B}$, there exists always a Klein 4-group $V_4 \leq PSL(2, q)$, which fixes B by the definition of Steiner 4-designs, and some additional point $x \in X$. We will distinguish two cases according as x is incident with B or not and examine for each case the list of possible subgroups of $PSL(2, q)$ with their orbits on the projective line (cf. Lemmas 8.3-8.11). Let $x \in B$. Then, clearly $k \equiv 1 \pmod 4$. It follows that we only have to consider the subcase when $PSL(2, q)_B$ is conjugate to $PSL(2, \bar{q})$ with $\bar{q}^m = q$, $m \geq 1$. In view of Lemma 8.7, we conclude then that $k = \bar{q} + 1$. By condition (B), we have hence

$$(q - 2)\,|PSL(2, q)_{0B}| = \bar{q}(\bar{q} - 1)(\bar{q} - 2)s \qquad (8.18)$$

$$\text{with } |PSL(2, q)_{0B}| = \bar{q}(\bar{q} - 1) \cdot \begin{cases} s, \text{ or} \\ 1. \end{cases}$$

Since $q = 2^{s^u}$, we can write $\bar{q} = 2^{s^w}$ for some integer $0 \leq w \leq u$, and $q = \bar{q}^m = \bar{q}^{s^{u-w}}$. As $k = \bar{q} + 1 = 2^{s^w} + 1 > 4$, it follows in particular that $w \geq 1$, and hence $s < 2^{s^w} = \bar{q}$. Thus, using equation (8.18), we obtain

$$\bar{q}^{s^{u-w}} - 2 = q - 2 \leq (\bar{q} - 2)s < \bar{q}^2 - 2s.$$

But, as clearly $u - w \geq 1$ (otherwise, $k = q + 1$, a contradiction to Corollary 1.17), this gives a contradiction for every prime s.

Now, let $x \notin B$. Then, clearly $k \equiv 0 \pmod 4$. We may restrict ourselves to the examination of the following subcases:

(i) $PSL(2, q)_B$ is conjugate to A_4 for $s = 2$, and $k = 12$ in view of Lemma 8.9.

(ii) $PSL(2, q)_B$ is conjugate to an elementary Abelian subgroup of order $\bar{q} \mid q$ of $PSL(2, q)$, and $k = \bar{q}$ due to Lemma 8.5.

(iii) $PSL(2, q)_B$ is conjugate to a semi-direct product of an elementary Abelian subgroup of order $\bar{q} \mid q$ with a cyclic subgroup of order c of $PSL(2, q)$ with $c \mid \bar{q} - 1$ and $c \mid q - 1$, and $k = \bar{q}$ or $\bar{q}c$ by Lemma 8.6.

(iv) $PSL(2, q)_B$ is conjugate to $PSL(2, \bar{q})$ with $\bar{q}^m = q$, $m \geq 1$, acting outside the $\bar{q} + 1$ points mentioned in the case where x has been incident with B, and Lemma 8.7 yields $k = \bar{q}(\bar{q} - 1)$ if m is even, or $k = (\bar{q} + 1)\bar{q}(\bar{q} - 1)$.

Again, we can write in the following $\bar{q} = 2^{s^w}$ for some integer $0 \leq w \leq u$, and $q = \bar{q}^m = \bar{q}^{s^{u-w}}$.

ad (i): Applying condition (B) gives

$$(q - 2)\,|PSL(2, q)_{0B}| = 11 \cdot 10 \cdot 9 \cdot 2$$

$$\text{with } |PSL(2, q)_{0B}| = \begin{cases} 2, \text{ or} \\ 1, \end{cases}$$

which is clearly impossible.

ad (iii): Let $k = \bar{q}$. By condition (B), we have

$$(q - 2)\,|PSL(2, q)_{0B}| = (\bar{q} - 1)(\bar{q} - 2)(\bar{q} - 3)s \tag{8.19}$$

$$\text{with } |PSL(2, q)_{0B}| = c \cdot \begin{cases} s, \text{ or} \\ 1. \end{cases}$$

As we may assume that $k = \bar{q} = 2^{s^w} > 4$, we have in particular $w \geq 1$, and hence $s < 2^{s^w} = \bar{q}$. Thus, using equation (8.19), we obtain

$$q - 2 = \bar{q}s^{u-w} - 2 < \bar{q}^3 s < \bar{q}^4.$$

Since clearly $u - w \geq 1$ (otherwise $k = q$, which is not possible by Corollary 1.17) this gives a contradiction for $s \geq 5$. If $s = 2$, then $\bar{q}^{2^{u-w}} - 2 < 2\bar{q}^3$ must hold, which cannot be true for $u - w > 1$. For $s = 3$, we may also assume that $u - w = 1$ since otherwise, we would have $q = \bar{q}^{3^{u-w}} \geq \bar{q}^9$, again a contradiction to the inequality above. As $c \mid \bar{q} - 1$, it follows for both cases from equation (8.19) that

$$\bar{q} - 2 \mid q - 2,$$

and hence

$$2^{s^w - 1} - 1 \mid 2^{s^u - 1} - 1.$$

Thus, clearly

$$s^w - 1 \mid s^u - 1$$

and

$$w \mid u.$$

Therefore, we may conclude that $w = 1$ and $u = 2$. For $s = 2$, it follows that $k = \bar{q} = 4$, which has been excluded. For $s = 3$, we have $\bar{q} = 8$ and $q = 512$, and equation (8.19) gives

$$510 \cdot |PSL(2, q)_{0B}| = 7 \cdot 6 \cdot 5 \cdot 3$$

$$\text{with } |PSL(2, q)_{0B}| = c \cdot \begin{cases} 3, \text{ or} \\ 1, \end{cases}$$

which is clearly impossible.

Now, let $k = \bar{q}c$. Then, condition (B) gives

$$(q - 2)\,|PSL(2, q)_{0B}| = (\bar{q}c - 1)(\bar{q}c - 2)(\bar{q}c - 3)s \tag{8.20}$$

$$\text{with } |PSL(2, q)_{0B}| = \begin{cases} s, \text{ or} \\ 1. \end{cases}$$

Polynomial division with remainder gives

$$2^{s^u-1} - 1 = \left(\sum_{i=1}^{\overline{m}} \frac{2^{s^u-1}}{(c \cdot 2^{s^w-1})^i} \right) \left(c \cdot 2^{s^w-1} - 1 \right) + \frac{2^{s^u-1}}{(c \cdot 2^{s^w-1})^{\overline{m}}} - 1$$

for a suitable $\overline{m} \in \mathbb{N}$ (such that

$$\deg\left(\frac{2^{s^u-1}}{(c \cdot 2^{s^w-1})^{\overline{m}}} - 1 \right) < \deg\left(c \cdot 2^{s^w-1} - 1 \right)$$

as is well-known). As c is odd, clearly $\left(\frac{2^{s^u-1}}{c \cdot 2^{s^w-1}} \right)^{\overline{m}} \neq 1$, and it follows that $\overline{q}c - 2$ does not divide $q - 2$, yielding a contradiction to equation (8.20).

ad (ii): Let $k = \overline{q}$. By condition (B), we have

$$(q - 2)\,|PSL(2,q)_{0B}| = (\overline{q} - 1)(\overline{q} - 2)(\overline{q} - 3)s$$

$$\text{with } |PSL(2,q)_{0B}| = \begin{cases} s, \text{ or} \\ 1. \end{cases}$$

As is easily seen, we may argue, mutatis mutandis, as in subcase (iii), $k = \overline{q}$.

ad (iv): If $m > 1$ even and $k = \overline{q}(\overline{q} - 1)$, then, in view of condition (B), we have

$$(q - 2)\,|PSL(2,q)_{0B}| = (\overline{q}^2 - \overline{q} - 1)(\overline{q}^2 - \overline{q} - 2)(\overline{q}^2 - \overline{q} - 3)s$$

$$\text{with } |PSL(2,q)_{0B}| = (\overline{q} + 1) \cdot \begin{cases} s, \text{ or} \\ 1. \end{cases}$$

As obviously $(\overline{q}^2 - \overline{q} - 1, \overline{q} + 1) = 1$, it follows that $\overline{q}^2 - \overline{q} - 1 \mid q - 2$ must hold, which is impossible as we have already seen via polynomial division (8.16) with remainder. For $k = \overline{q}^3 - \overline{q}$, condition (B) implies that

$$(q - 2)\,|PSL(2,q)_{0B}| = (\overline{q}^3 - \overline{q} - 1)(\overline{q}^3 - \overline{q} - 2)(\overline{q}^3 - \overline{q} - 3)s \qquad (8.21)$$

$$\text{with } |PSL(2,q)_{0B}| = \begin{cases} s, \text{ or} \\ 1. \end{cases}$$

We already know that $k = (\overline{q} + 1)\overline{q}(\overline{q} - 1) \equiv 0 \pmod{4}$, and thus $\overline{q} > 2$. If $|PSL(2,q)_{0B}| = s$, then

$$q = (\overline{q}^3 - \overline{q} - 1)(\overline{q}^3 - \overline{q} - 2)(\overline{q}^3 - \overline{q} - 3) + 2 = \overline{q}^9 - l$$

$$\text{with } l = 3\overline{q}^7 + 6\overline{q}^6 - 3\overline{q}^5 - 12\overline{q}^4 - 10\overline{q}^3 + 6\overline{q}^2 + 11\overline{q} + 4.$$

As clearly $l > 0$, we have $q < \overline{q}^9$. On the other hand, for $\overline{q} > 2$ certainly $l < \overline{q}^8(\overline{q}-1)$ and hence $q > \overline{q}^8$ must hold, a contradiction to the fact that $q = \overline{q}^m$ for some $m \geq 1$. If $|PSL(2,q)_{0B}| = 1$, then equation (8.21) gives

$$q = ls + 2 \text{ with } l = (\overline{q}^3 - \overline{q} - 1)(\overline{q}^3 - \overline{q} - 2)(\overline{q}^3 - \overline{q} - 3)$$
$$= \overline{q}^9 - 3\overline{q}^7 - 6\overline{q}^6 + 3\overline{q}^5 + 12\overline{q}^4 + 10\overline{q}^3 - 6\overline{q}^2 - 11\overline{q} - 6.$$

Since $\bar{q} = 2^{s^w} > 2$, we conclude that $w \geq 1$ and $s < 2^{s^w} = \bar{q}$. As obviously $l < \bar{q}^9 - 1$, it follows $q < (\bar{q}^9 - 1)\bar{q} + 2 < \bar{q}^{10}$. On the other hand, for $\bar{q} > 2$ clearly $q = ls + 2 \geq 2(l + 1) > \bar{q}^9$ must hold, which again is impossible.

Case (2b): $N = PSL(d, q)$, $d \geq 3$, $v = \frac{q^d - 1}{q - 1}$.

We have here $\mathrm{Aut}(N) = P\Gamma L(d, q) \rtimes \langle \iota_\beta \rangle$, where ι_β denotes the graph automorphism from Chapter 6. In the following, let $n = (d, q - 1)$.

Let us first assume that $d = 3$. In order to show that G with $PSL(3, q)$ as simple normal subgroup cannot act on any non-trivial 4-$(q^2 + q + 1, k, 1)$ design, we prove first that $k \leq q + 1$ as an upper bound for the block size k must hold. For any line $\mathcal{G}$ in the underlying projective plane $PG(2, q)$, the translation group $T(\mathcal{G})$ operates regularly on the points of $PG(2, q) \backslash \mathcal{G}$ and acts trivially on $\mathcal{G}$. Thus, $T(\mathcal{G})$ fixes a block $B \in \mathcal{B}$ if four or more distinct points of B lie on $\mathcal{G}$. By the definition of Steiner 4-designs, we may choose in $PG(2, q)$ four distinct collinear points $x_1, x_2, x_3, x_4 \in X$, which are incident with a unique block $B \in \mathcal{B}$. Let $\mathcal{G}$ denote the line of $PG(2, q)$ through $x_1, x_2, x_3, x_4 \in X$. Consequently, if the block B contains at least one further point of $PG(2, q) \backslash \mathcal{G}$, then it must contain all points of $PG(2, q) \setminus \mathcal{G}$, thus at least $q^2 + 4$ many, which is not possible as $k \leq \lfloor \frac{v}{5} + 3 \rfloor$ by Proposition 1.16 (a). Therefore, B is completely contained in $\mathcal{G}$, and hence $k \leq q + 1$.

Now, by the definition of Steiner 4-designs, we may consider a 4-subset consisting of three distinct collinear points $x_1, x_2, x_3 \in X$ and one non-collinear point $x_4 \in X$, which is incident with a unique block $B \in \mathcal{B}$. If B contains a fourth point on the line $\mathcal{G}$ of $PG(2, q)$ through $x_1, x_2, x_3 \in X$, then by the same arguments as above using the translation group $T(\mathcal{G})$, we conclude that B lies completely in $\mathcal{G}$, a contradiction. Thus, we may assume that B contains only further points which are not on $\mathcal{G}$. Without restriction, we may identify $x_1 = \langle (1,0,0) \rangle$, $x_2 = \langle (0, 0, 1) \rangle$, $x_3 \in \langle x_1, x_2 \rangle$, and $x_4 = \langle (0, 1, 0) \rangle$. As is known, the cyclic group

$$\left\{ \begin{pmatrix} c & & \\ & c^{-2} & \\ & & c \end{pmatrix} \middle| c \in GF(q)^* \right\}$$

of linear transformations on the associated vector space $V = V(3, q)$ induces a group U of dilatations of order $\frac{q-1}{n}$ on $PG(2, q)$ with axis the line $\mathcal{G} = \langle x_1, x_2 \rangle$ and as center the point x_4. It is clear that U fixes each point of its axis as well as its center. Furthermore, U acts semi-regularly on the points of $PG(2, q) \setminus (\mathcal{G} \cup \{x_4\})$, and hence all point-orbits on $PG(2, q) \setminus (\mathcal{G} \cup \{x_4\})$ have length $\frac{q-1}{n}$. As U fixes each of the points $x_1, x_2, x_3, x_4 \in X$ and hence in particular B, we get

$$k \equiv 4 \left(\mathrm{mod}\ \frac{q-1}{n} \right).$$

Due to the fact that $k \leq q + 1$, this is obviously impossible if $3 \nmid q - 1$, and for $3 \mid q - 1$, we conclude that

$$k = \frac{q-1}{3} + 4 \text{ or } k = 2 \cdot \frac{q-1}{3} + 4. \tag{8.22}$$

If we assume that $q > 7$, then indeed $q \geq 13$, and hence we obtain $\frac{q-1}{3} \geq 4$, which means that we have at least four distinct collinear points on some line $\mathcal{H}$ of $PG(2, q)$, and we may argue as above using the translation group $T(\mathcal{H})$ that then B lies completely in $\mathcal{H}$, which is obviously impossible. Therefore, we only have to consider the cases when $q = 4$ or 7. For $q = 7$, condition (8.22) implies $k = 6$ or 8, whereas $k = 6$ can immediately be ruled out using Lemma 1.14 (c). If any 4-$(57, 8, 1)$ design exists, then there must also exist a derived 3-$(56, 7, 1)$ design. But, for $t = 3$, it follows from Lemma 1.14 (c) that then in particular 54 must be divisible by 5, a contradiction. Now, let us assume that $q = 4$. Then only $k = 5$ can occur. We have the situation of two intersecting lines $\mathcal{G}$ and $\mathcal{H}$, and we may distinguish the two cases according to whether or not their intersecting point $x \in \mathcal{G} \cap \mathcal{H}$ is incident with B. In the first case, $\mathcal{G}$ and $\mathcal{H}$ are precisely the lines which intersect B in exactly three distinct points. We will show that then $|PSL(3, 4)_B| \leq 8$. Let B be fixed under $PSL(3, 4)$. Thus, the set consisting of the lines $\mathcal{G}$ and $\mathcal{H}$ is also fixed. Hence, $PSL(3, 4)_B$ has a normal subgroup U of index at most 2 which fixes both lines. Then U also fixes their intersecting point x and the remaining 2-subset of B on each line. Thus, U has a normal subgroup U_1 of index at most 2 which fixes pointwise any of the two 2-subsets and furthermore U_1 has a normal subgroup U_2 of index at most 2 which fixes pointwise both 2-subsets. Since in $PSL(3, q)$ only the identity element fixes pointwise some non-degenerate quadrangle, the claim follows. In the second case there is exactly one line which intersects B in exactly three distinct points, and by similar arguments it can be verified that here $|PSL(3, 4)_B| \leq 2$. However, as there are 21 projective lines as blocks in $PG(2, 4)$, it follows that $|PSL(3, 4)_B| \geq \frac{|PSL(3,4)|}{b-21} = \frac{20160}{1176} > 17$, a contradiction in both cases. Thus $PSL(3, q)$, and hence also G with $PSL(3, q)$ as simple normal subgroup, cannot act on any non-trivial 4-$(q^2 + q + 1, k, 1)$ design.

Now, we consider the case when $d > 3$. Via induction over d, we will verify that $G \leq \mathrm{Aut}(\mathcal{D})$ cannot act on any non-trivial Steiner 4-design $\mathcal{D}$. For this, let us assume that there is a counterexample with d minimal. Without restriction, we can choose four distinct points x_1, x_2, x_3, x_4 from a hyperplane $\mathcal{H}$ of $PG(d - 1, q)$. Analogously as above, it can be shown that the unique block $B \in \mathcal{B}$ which is incident with the 4-subset $\{x_1, x_2, x_3, x_4\}$ is contained completely in $\mathcal{H}$. Thus, $\mathcal{H}$ induces a 4-$(\frac{q^{d-1}-1}{q-1}, k, 1)$ design, on which G containing $PSL(d - 1, q)$ operates as simple normal subgroup. Inductively, we obtain the minimal counter-example for $d = 3$, which is impossible as shown above.

Case (3): $N = PSU(3, q^2)$, $v = q^3 + 1$, $q = p^e > 2$.

Here $\mathrm{Aut}(N) = P\Gamma U(3, q^2)$, and $|G| = (q^3 + 1)q^3 \frac{(q^2-1)}{n} a$ with $n = (3, q + 1)$ and $a \mid 2ne$. For the existence of flag-transitive Steiner 4-designs, necessarily

$$r = \frac{q^3(q^3 - 1)(q^3 - 2)}{(k - 1)(k - 2)(k - 3)} \,\bigg|\, |G_0| \,\bigg|\, |P\Gamma U(3, q^2)_0| = q^3(q^2 - 1)2e$$

must hold in view of Lemma 4.1. As obviously $(q^2 + q + 1, q + 1) = 1$ and

$(q^3 - 2, q + 1) = (3, q + 1) = n$, we have in particular

$$(q^3 - 2)(q^2 + q + 1) \mid (k - 1)(k - 2)(k - 3)2ne, \text{ where } e \leq \log_2 q. \tag{8.23}$$

On the other hand, Corollary 1.17 yields $k \leq \lfloor \sqrt{q^3 + 1} + \frac{5}{2} \rfloor < q^{\frac{3}{2}} + 3$. Hence, using property (8.23), we have only a small number of possibilities to check, which can easily be eliminated by hand.

Case (4): $N = Sz(q)$, $v = q^2 + 1$, $q = 2^{2e+1} > 2$.

We have $\mathrm{Aut}(N) = Sz(q) \rtimes \langle \alpha \rangle$, where α denotes the Frobenius automorphism $GF(q) \longrightarrow GF(q)$, $x \mapsto x^2$. Thus, by Dedekind's law, $G = Sz(q) \rtimes (G \cap \langle \alpha \rangle)$, and $|G| = (q^2 + 1)q^2(q - 1)a$ with $a \mid 2e + 1$. From Remark 4.15, we hence obtain

$$(q^2 - 2)(q + 1) = (k - 1)(k - 2)(k - 3) \frac{a}{|G_{xB}|} \text{ if } x \in B.$$

First, we show that every element $g \in G$ that fixes three distinct points must fix at least five distinct points. Let us assume that $g \in G$ with $|\mathrm{Fix}_X(g)| \geq 3$. Let $x \in \mathrm{Fix}_X(g)$, and P the normal Sylow 2-subgroup of $Sz(q)_x$ acting regularly on $X \setminus \{x\}$. Furthermore, let $\mathcal{C}_P(g)$ denote the centralizer of g in P, where clearly $\mathcal{C}_P(g) = P \cap \mathcal{C}_{Sz(q)}(g)$. If $y, z \in \mathrm{Fix}_X(g) \setminus \{x\}$, then $z = y^h$ with $h \in P$. Thus, as $y^{hg} = y^h = y^{gh}$, we conclude that

$$[h^{-1}, g^{-1}] \in G_{xy} \cap [P, G_x] \leq P_y = 1.$$

Then $h \in \mathcal{C}_P(g)$, and hence $\mathcal{C}_P(g)$ acts point-transitively on $\mathrm{Fix}_X(g) \setminus \{x\}$. Therefore, as $|\mathrm{Fix}_X(g)| \geq 3$, it follows that $|\mathrm{Fix}_X(g)| \equiv 1 \pmod 2$. Clearly, the set $\mathrm{Fix}_X(g)$ is left invariant by $\mathcal{C}_{Sz(q)}(g)$ and $\mathcal{C}_{Sz(q)}(g)$ operates on $\mathrm{Fix}_X(g)$. Since $x \in \mathrm{Fix}_X(g)$ can be chosen arbitrarily, it follows that $\mathcal{C}_{Sz(q)}(g)$ operates point-transitively on $\mathrm{Fix}_X(g)$, and thus $|\mathrm{Fix}_X(g)| \mid |Sz(q)|$. As the order of $Sz(q)$ is not divisible by 3, clearly $|\mathrm{Fix}_X(g)| \neq 3$, and due to the fact that $|\mathrm{Fix}_X(g)| \equiv 1 \pmod 2$, we have $|\mathrm{Fix}_X(g)| \geq 5$.

Since G is block-transitive, it is sufficient to consider some appropriate, unique block $B \in \mathcal{B}$. As clearly $\langle \alpha \rangle \leq \mathrm{Aut}(N)_{0,1,\infty}$, it follows from above that $\langle \alpha \rangle$ must fix some fourth point, and hence $G \cap \langle \alpha \rangle \leq G_{0B}$ by the definition of Steiner 4-designs. Thus, we have particularly

$$(q^2 - 2)(q + 1) \leq (k - 1)(k - 2)(k - 3),$$

which does obviously not hold for $k \leq q + 2$. On the other hand, Corollary 1.17 implies $k \leq \lfloor \sqrt{q^2 + 1} + \frac{5}{2} \rfloor < q + 3$, a contradiction.

Case (5): $N = Re(q)$, $v = q^3 + 1$, $q = 3^{2e+1} > 3$.

Here $\mathrm{Aut}(N) = Re(q) \rtimes \langle \alpha \rangle$, where α denotes the Frobenius automorphism $GF(q) \longrightarrow GF(q)$, $x \mapsto x^3$. Thus, by Dedekind's law, $G = Re(q) \rtimes (G \cap \langle \alpha \rangle)$, and $|G| = (q^3 + 1)q^3(q - 1)a$ with $a \mid 2e + 1$. It follows from Remark 4.15 that

$$(q^3 - 2)(q^2 + q + 1) = (k - 1)(k - 2)(k - 3) \frac{a}{|G_{xB}|} \text{ if } x \in B.$$

First, we show that every element $g \in G$ that fixes three distinct points must also fix a fourth point. Let us assume that $g \in G$ with $|\text{Fix}_X(g)| \geq 3$. Let $x \in \text{Fix}_X(g)$, and P the normal Sylow 3-subgroup of $Re(q)_x$ acting regularly on $X \setminus \{x\}$. As in Case (4), it can be shown that then $\mathcal{C}_P(g)$ acts point-transitively on $\text{Fix}_X(g) \setminus \{x\}$. Thus, we have $|\text{Fix}_X(g)| \equiv 0 \pmod 2$, and the claim follows.

Since G is block-transitive, it is sufficient to consider some appropriate, unique block $B \in \mathcal{B}$. As clearly $\langle \alpha \rangle \leq \text{Aut}(N)_{0,1,\infty}$, we deduce from the above that $G \cap \langle \alpha \rangle \leq G_{0B}$ by the definition of Steiner 4-designs. Hence, we have in particular

$$(q^3 - 2)(q^2 + q + 1) \leq (k-1)(k-2)(k-3),$$

which is not possible as Corollary 1.17 yields $k \leq \lfloor \sqrt{q^3 + 1} + \frac{5}{2} \rfloor < q^{\frac{3}{2}} + 3$.

Case (6): $N = Sp(2d, 2)$, $d \geq 3$, $v = 2^{2d-1} \pm 2^{d-1}$.

As here $|\text{Out}(N)| = 1$, we have $N = G$. Let X^+, respectively, X^- denote the set of points on which G operates. It is well-known that G_z acts on $X^\pm \setminus \{z\}$ as $O^\pm(2d, 2)$ does in its usual rank 3 manner on singular points of the underlying non-degenerate orthogonal space $V^\pm = V^\pm(2d, 2)$.

It is easily seen that there are $2^{2d-2}(2^d \mp 1)(2^{d-1} \pm 1)$ hyperbolic pairs in $V^\pm$, and by Witt's theorem, $O^\pm(2d, 2)$ is transitive on these hyperbolic pairs (cf. [73, Chap. II, Thm. 9.13]). Let $\{x, y\}$ denote a hyperbolic pair, and $\mathcal{E} = \langle x, y \rangle$ the hyperbolic plane spanned by $\{x, y\}$. As $\mathcal{E}$ is non-degenerate, we have the orthogonal decomposition

$$V^\pm = \mathcal{E} \perp \mathcal{E}^\perp.$$

Clearly, $O^\pm(2d, 2)_{\{x,y\}}$ stabilizes $\mathcal{E}^\perp$ as a subspace, which implies that $O^\pm(2d, 2)_{\{x,y\}} \cong O^\pm(2d - 2, 2)$. Therefore, we have

$$O^\pm(2d - 2, 2) \cong O^\pm(2d, 2)_{\{x,y\}} \trianglelefteq O^\pm(2d, 2)_{\mathcal{E}} = G_{z,\mathcal{E}}.$$

Since $O^\pm(2d - 2, 2)$ acts transitively on the singular points of the $(2d - 2)$-dimensional orthogonal subspace, we conclude that the smallest orbit on $V^\pm \setminus \mathcal{E}$ under $G_{z,\mathcal{E}}$ has length at least $2^{2d-3} \pm 2^{d-2}$. If the unique block $B \in \mathcal{B}$ which is incident with the 4-subset $\{x, y, x+y, z\}$ contains some singular point in $V^\pm \setminus \mathcal{E}$, then we would have $k \geq 2^{2d-3} \pm 2^{d-2} + 4$, a contradiction to Corollary 1.17. Thus, all points of B apart from z lie completely in $\mathcal{E}$. By the flag-transitivity of G, it follows that for each block all points apart from a singleton must be contained in an affine plane. Thus $k = 5$, which is impossible since $k \equiv 0 \pmod 4$ by Lemma 1.14 (c).

Case (7): $N = PSL(2, 11)$, $v = 11$.

As is known, this exceptional permutation action occurs inside the Mathieu group M_{24} in its action on 24 points. This set can be partitioned into two sets X_1 and X_2 of 12 points each, such that the setwise stabilizer of X_1 is the Mathieu group M_{12}. The stabilizer in this latter group of a point x in X_1 is isomorphic to M_{11} and operates (apart from its natural 4-transitive action on $X_1 \setminus \{x\}$)

3-transitively on the 12 points of X_2. The one-point stabilizer in this action of degree 12 is $PSL(2, 11)$ acting 2-transitively on 11 points. The geometry preserved by the 3-transitive action of M_{11} is not a Steiner t-design, but a 3-$(12, 6, 2)$ design (e.g. [8, Chap. IV, 5.3]). Thus, the derived design $\mathcal{D}$ on which $G \leq \text{Aut}(\mathcal{D})$ acts cannot be a Steiner design.

Case (9): $N = M_v$, $v = 11, 12, 22, 23, 24$.

If $v = 11, 12, 23$ or 24, then $G = M_v$ is always 4-transitive, and thus Theorem 5.2 gives the designs described in Theorem 8.1. Obviously, flag-transitivity holds as the 4-transitivity of G implies that G_x acts block-transitively on the derived Steiner 3-design $\mathcal{D}_x$ for any $x \in X$. For $v = 22$, Corollary 1.17 gives $k \leq 7$, and again the cases for k can easily be eliminated by Lemma 1.14 (c).

Case (10): $N = M_{11}$, $v = 12$.

As already illustrated in Case (7), $G \leq \text{Aut}(\mathcal{D})$ cannot act on any Steiner design $\mathcal{D}$.

This completes the proof of Theorem 8.1.   □

Chapter 9

The Classification of Flag-transitive Steiner 5-Designs

9.1 Introduction

In this chapter, we present the complete classification of all flag-transitive Steiner 5-designs. The result relies on the classification of the finite 3-homogeneous permutation groups, which itself depends on the finite simple group classification (see Chapter 2). With regard to Section 2.2, we may consider two types of 3-homogeneous permutation groups. Specifically, as in the previous chapter the groups of almost simple type with the projective group $PSL(2, q)$ as simple normal subgroup need extensive consideration.

9.2 Main Result

The classification of all non-trivial Steiner 5-designs admitting a flag-transitive group of automorphisms can be stated as follows:

Theorem 9.1. *Let $\mathcal{D} = (X, \mathcal{B}, I)$ be a non-trivial Steiner 5-design. Then $G \leq \mathrm{Aut}(\mathcal{D})$ acts flag-transitively on $\mathcal{D}$ if and only if one of the following occurs:*

(1) $\mathcal{D}$ *is isomorphic to the Mathieu-Witt 5-$(12, 6, 1)$ design, and $G \cong M_{12}$,*

(2) $\mathcal{D}$ *is isomorphic to the Mathieu-Witt 5-$(24, 8, 1)$ design, and $G \cong PSL(2, 23)$ or $G \cong M_{24}$.*

Remark 9.2. We note that in Part (2), $G \cong PSL(2, 23)$ acts sharply flag-transitively on $\mathcal{D}$, and furthermore that M_{24} as the full group of automorphisms of $\mathcal{D}$ contains only one conjugacy class of subgroups isomorphic to $PSL(2, 23)$ (cf. [35]).

9.3   Groups of Automorphisms of Affine Type

In this section, we start with the proof of Theorem 9.1. Let $\mathcal{D} = (X, \mathcal{B}, I)$ be a non-trivial Steiner 5-design with $G \leq \mathrm{Aut}(\mathcal{D})$ acting flag-transitively on $\mathcal{D}$ throughout this chapter. We recall that due to Proposition 4.14, we may restrict ourselves to the consideration of the finite 3-homogeneous permutation groups listed in Section 2.2. Clearly, in the following we may assume that $k > 5$ as trivial Steiner 5-designs are excluded. Let us first assume that G is of affine type.

   Case (1): $G \cong AGL(1, 8)$, $A\Gamma L(1, 8)$ or $A\Gamma L(1, 32)$.

   We may assume that $k > 5$. For $v = 8$, we obtain $k = 6$ by Corollary 1.17, which is not possible in view of Lemma 1.14 (b). If $v = 32$, then $|G| = 5v(v - 1)$, and Lemma 4.1 immediately yields that $G \leq \mathrm{Aut}(\mathcal{D})$ cannot act flag-transitively on any non-trivial Steiner 5-design $\mathcal{D}$.

   Case (2): $G_0 \cong SL(d, 2)$, $d \geq 2$.

   Let e_i denote the i-th standard basis vector of the vector space $V = V(d, 2)$, and $\langle e_i \rangle$ the 1-dimensional vector subspace spanned by e_i.
   We assume that $v = 2^d > k > 5$. For $d = 3$, we have $v = 8$ and $k = 6$ by Corollary 1.17, which is not possible in view of Lemma 1.14 (b) again. So, we may assume that $d > 3$. We remark that clearly any five distinct points are non-coplanar in $AG(d, 2)$ and hence generate an affine subspace of dimension at least 3. Let $\mathcal{E} = \langle e_1, e_2, e_3 \rangle$ denote the 3-dimensional vector subspace spanned by e_1, e_2, e_3. Then $SL(d, 2)_\mathcal{E}$ and therefore also $G_{0, \mathcal{E}}$ acts point-transitively on $V \setminus \mathcal{E}$. If the unique block $B \in \mathcal{B}$ which is incident with the 5-subset $\{0, e_1, e_2, e_3, e_1 + e_2\}$ contains some point outside $\mathcal{E}$, then it would already contain all points of $V \setminus \mathcal{E}$, and hence $k \geq 2^d - 8 + 5 = 2^d - 3$, which is not possible in view of Corollary 1.17. Thus, B lies completely in $\mathcal{E}$, and by the flag-transitivity of G, it follows that each block must be contained in a 3-dimensional affine subspace. Then clearly $k \leq 8$. On the other hand, for $\mathcal{D}$ to be a block-transitive 5-design admitting $G \leq \mathrm{Aut}(\mathcal{D})$, we obtain from [1] the necessary (and sufficient) condition that $2^d - 3$ must divide $\binom{k}{4}$, and hence it follows for each respective value of k that $d = 3$, contradicting our assumption.

   Case (3): $G_0 \cong A_7$, $v = 2^4$.

   Since $v = 2^4$, we obtain from Corollary 1.17 that $k \leq 7$. But, Lemma 4.1 easily rules out the cases when $k = 6$ or 7.

9.4   Groups of Automorphisms of Almost Simple Type

We consider in this section successively those cases where G is of almost simple type. For Case (2), Lemmas 8.2-8.11 from Section 8.4 will be required.

Case (1): $N = A_v$, $v \geq 5$.

We may assume that $v \geq 7$. But then A_v, and hence also G, is 5-transitive and does not act on any non-trivial Steiner 5-design $\mathcal{D}$ in view of Theorem 5.2.

Case (2): $N = PSL(2, q)$, $v = q + 1$, $q = p^e > 3$.

Here $\mathrm{Aut}(N) = P\Gamma L(2, q)$, and $|G| = (q + 1)q\frac{(q-1)}{n}a$ with $n = (2, q - 1)$ and $a \mid ne$. We may assume that $q \geq 5$. We will show that only the flag-transitive design given in Part (2) of Theorem 9.1 with $G \cong PSL(2, 23)$ can occur.

We will first assume that $N = G$. Then, by Remark 4.15, we obtain

$$(q - 2)(q - 3)\,|PSL(2, q)_{0B}| \cdot n = (k - 1)(k - 2)(k - 3)(k - 4). \tag{9.1}$$

In view of Proposition 1.16 (b), we have

$$q - 3 \geq (k - 3)(k - 4), \tag{9.2}$$

and thus it follows from equation (9.1) that

$$(q - 2)\,|PSL(2, q)_{0B}| \cdot n \leq (k - 1)(k - 2). \tag{9.3}$$

If we assume that $k \geq 9$, then clearly

$$(k - 1)(k - 2) < 2(k - 3)(k - 4),$$

and hence we obtain

$$(q - 2)\,|PSL(2, q)_{0B}| \cdot n < 2(q - 3)$$

due to Proposition 1.16 (b) again, which is obviously only possible when $|PSL(2, q)_{0B}| \cdot n = 1$. Thus, in particular q has to be even. But then, considering equation (9.1) gives that the left-hand side of the equation is not divisible by 4, whereas obviously the right-hand side is always divisible by 8, a contradiction. If $k < 9$, then, using equation (9.1) and inequality (9.2), the very few remaining possibilities for k can immediately be ruled out by hand, except for the case when $k = 8$, $q = 23$ and $|PSL(2, q)_{0B}| = 1$. Referring to Chapter 4, Example 4.6, there exists (up to isomorphism) for the parameters $t = 5$, $v = 24$ and $k = 8$ only the Mathieu-Witt 5-(24, 8, 1) design $\mathcal{D}$, which can alternatively be constructed from $PSL(2, 23)$ in its natural 3-homogeneous action on the elements of $GF(23) \cup \{\infty\}$. Furthermore, it can be shown that the setwise stabilizer $PSL(2, 23)_B$ of an appropriate, unique block $B \in \mathcal{B}$ is a dihedral group of order 8 (see, e.g., [8, Chap. IV, 1.5], [31, Chap. XIV, 115], and [126, Thm. 5] for a uniqueness proof). Thus, using Lemma 1.14 (b), we obtain $b = 759 = [PSL(2, 23) : PSL(2, 23)_B]$, and hence $PSL(2, 23)$ acts block-transitively on $\mathcal{D}$. As for $q = 23$, the dihedral group of order 8 has only orbits of length 8 in view of Lemma 8.4 (ii)(a), clearly $PSL(2, 23)_B$ acts transitively on the points of B. Since we have $|PSL(2, 23)_{0B}| = 1$, it follows that $PSL(2, 23)$ acts even sharply flag-transitively on $\mathcal{D}$.

Now, let us assume that $N < G \le \mathrm{Aut}(N)$. We recall that $q = p^e > 3$, and will distinguish in the following the cases $p > 3$, $p = 2$, and $p = 3$.

First, let $p > 3$. We define

$$G^* = G \cap (PSL(2, q) \rtimes \langle \tau_\alpha \rangle)$$

with $\tau_\alpha \in \mathrm{Sym}(GF(p^e) \cup \{\infty\}) \cong S_v$ of order e induced by the Frobenius automorphism $\alpha : GF(p^e) \longrightarrow GF(p^e)$, $x \mapsto x^p$. Then, by Dedekind's law, we can write

$$G^* = PSL(2, q) \rtimes (G^* \cap \langle \tau_\alpha \rangle). \tag{9.4}$$

Defining $P\Sigma L(2, q) = PSL(2, q) \rtimes \langle \tau_\alpha \rangle$, it can easily be calculated that $P\Sigma L(2, q)_{0,1,\infty} = \langle \tau_\alpha \rangle$, and $\langle \tau_\alpha \rangle$ has precisely $p + 1$ distinct fixed points (cf., e.g., [43, Chap. 6.4, Lemma 2]). As $p > 3$, we conclude therefore that $G^* \cap \langle \tau_\alpha \rangle \le G^*_{0B}$ for some appropriate, unique block $B \in \mathcal{B}$ by the definition of Steiner 5-designs. Furthermore, clearly $PSL(2, q) \cap (G^* \cap \langle \tau_\alpha \rangle) = 1$. Hence, we have

$$
\begin{aligned}
\left| (0, B)^{G^*} \right| &= [G^* : G^*_{0B}] \\
&= [PSL(2, q) \rtimes (G^* \cap \langle \tau_\alpha \rangle) : PSL(2, q)_{0B} \rtimes (G^* \cap \langle \tau_\alpha \rangle)] \\
&= [PSL(2, q) : PSL(2, q)_{0B}] \\
&= \left| (0, B)^{PSL(2,q)} \right|.
\end{aligned}
\tag{9.5}
$$

Thus, if we assume that $G^* \le \mathrm{Aut}(\mathcal{D})$ acts already flag-transitively on $\mathcal{D}$, then we obtain $\left| (0, B)^{G^*} \right| = \left| (0, B)^{PSL(2,q)} \right| = bk$ in view of Remark 4.15. Hence, $PSL(2, q)$ must also act flag-transitively on $\mathcal{D}$, and we may proceed as in the case when $N = G$. Therefore, let us assume that $G^* \le \mathrm{Aut}(\mathcal{D})$ does not act flag-transitively on $\mathcal{D}$. Then, we conclude that $[G : G^*] = 2$ and G^* has exactly two orbits of equal length on the set of flags. Thus, by equation (9.5), we obtain for the orbit containing the flag $(0, B)$ that $\left| (0, B)^{G^*} \right| = \left| (0, B)^{PSL(2,q)} \right| = \frac{bk}{2}$. As $PSL(2, q)$ is normal in G, we have under $PSL(2, q)$ also precisely one further orbit of equal length on the set of flags. Then, proceeding similarly to the case $N = G$ for each orbit on the set of flags, we obtain (representative for the orbit containing the flag $(0, B)$) that

$$\frac{(q-2)(q-3)\,|PSL(2, q)_{0B}| \cdot n}{2} = (k-1)(k-2)(k-3)(k-4), \tag{9.6}$$

and as here $n = 2$, this is equivalent to

$$
\begin{aligned}
(q-2)(q-3)\,|PSL(2, q)_{0B}| &= (k-1)(k-2)(k-3)(k-4) \\
&= k(k^3 - 10k^2 + 35k - 50) + 24.
\end{aligned}
\tag{9.7}
$$

Hence, we have in particular

$$k \mid (q-2)(q-3)\,|PSL(2, q)_{0B}| - 24. \tag{9.8}$$

Since $PSL(2,q)_B$ can have one or two orbits of equal length on the points of B, we have

$$k \text{ or } \frac{k}{2} = \left|0^{PSL(2,q)_B}\right| = [PSL(2,q)_B : PSL(2,q)_{0B}]. \tag{9.9}$$

By the same arguments as in case $N = G$, we deduce from equation (9.7) that

$$(q-2)\,|PSL(2,q)_{0B}| \le (k-1)(k-2), \tag{9.10}$$

and assuming that $k \ge 9$, we obtain

$$(q-2)\,|PSL(2,q)_{0B}| < 2(q-3),$$

which is clearly only possible when $|PSL(2,q)_{0B}| = 1$. Hence, it follows that

$$(q-2)(q-3) = (k-1)(k-2)(k-3)(k-4), \tag{9.11}$$

and $k \mid (q-2)(q-3) - 24$ in view of property (9.8). On the other hand, for $k \ge 9$, we obtain from equation (9.9) that k or $\frac{k}{2} = |PSL(2,q)_B| \mid |PSL(2,q)| = \frac{q^3-q}{2}$, and thus in particular $k \mid q^3 - q$. As $(q^3 - q, (q-2)(q-3) - 24) \mid 2^3 \cdot 3 \cdot 11$, we have only a small number of possibilities for k to check, all of which can easily be ruled out by hand using equation (9.11). For $k < 9$, the very few remaining possibilities for k can immediately be ruled out by hand using inequality (9.2) and equation (9.7), except for the case when $k = 8$, $q = 23$ and $|PSL(2,q)_{0B}| = 2$. But, as involutions are fixed point free on the points of the projective line for $q \equiv 3 \pmod 4$ in view of Lemma 8.2, this is impossible.

Now, let $p = 2$. Then, clearly $N = PSL(2,q) = PGL(2,q)$, and we have $\mathrm{Aut}(N) = P\Sigma L(2,q)$. If we assume that $\langle \tau_\alpha \rangle \le P\Sigma L(2,q)_{0B}$ for some appropriate, unique block $B \in \mathcal{B}$, then, using the terminology of (9.4), we have $G^* = G = P\Sigma L(2,q)$ and as clearly $PSL(2,q) \cap \langle \tau_\alpha \rangle = 1$, we can apply equation (9.5). Thus, $PSL(2,q)$ must also be flag-transitive, which has already been considered. Therefore, we may assume that $\langle \tau_\alpha \rangle \not\le P\Sigma L(2,q)_{0B}$. Let s be a prime divisor of $e = |\langle \tau_\alpha \rangle|$. As the normal subgroup $H := (P\Sigma L(2,q)_{0,1,\infty})^s \le \langle \tau_\alpha \rangle$ of index s has precisely $p^s + 1$ distinct fixed points (see, e.g., [43, Chap. 6.4, Lemma 2]), we have, by the definition of Steiner 5-designs, $G \cap H \le G_{0B}$ for some appropriate, unique block $B \in \mathcal{B}$. It can then be deduced that $e = s^u$ for some $u \in \mathbb{N}$, since if we assume for $G = P\Sigma L(2,q)$ that there exists a further prime divisor $\bar{s}$ of e with $\bar{s} \ne s$, then $\overline{H} := (P\Sigma L(2,q)_{0,1,\infty})^{\bar{s}} \le \langle \tau_\alpha \rangle$ and H are both subgroups of $P\Sigma L(2,q)_{0B}$ by the flag-transitivity of $P\Sigma L(2,q)$, and hence $\langle \tau_\alpha \rangle \le P\Sigma L(2,q)_{0B}$, a contradiction. Furthermore, as $\langle \tau_\alpha \rangle \not\le P\Sigma L(2,q)_{0B}$, we may, by applying Dedekind's law, assume that

$$G_{0B} = PSL(2,q)_{0B} \rtimes (G \cap H).$$

Thus, by Remark 4.15, we obtain

$$(q-2)(q-3)\,|PSL(2,q)_{0B}|\,|G \cap H| = (k-1)(k-2)(k-3)(k-4)\,|G \cap \langle \tau_\alpha \rangle|.$$

Using that $k = \left|0^{G_B}\right| = [G_B : G_{0B}]$, we have more precisely

(A) if $G = PSL(2, q) \rtimes (G \cap H)$:

$$(q - 2)(q - 3)\,|PSL(2, q)_{0B}| = (k - 1)(k - 2)(k - 3)(k - 4)$$

$$\text{with } |PSL(2, q)_{0B}| = \frac{|PSL(2, q)_B|}{k}, \text{ or}$$

(B) if $G = P\Sigma L(2, q)$:

$$(q - 2)(q - 3)\,|PSL(2, q)_{0B}| = (k - 1)(k - 2)(k - 3)(k - 4)s$$

$$\text{with } |PSL(2, q)_{0B}| = \frac{|PSL(2, q)_B|}{k} \cdot \begin{cases} s, & \text{if } G_B = PSL(2, q)_B \rtimes \langle \tau_\alpha \rangle, \\ 1, & \text{if } G_B = PSL(2, q)_B \rtimes H. \end{cases}$$

As far as condition (A) is concerned, we may argue exactly as in the earlier case $N = G$. Thus, only condition (B) has to be examined, and we will also show that here $G \leq \mathrm{Aut}(\mathcal{D})$ cannot act flag-transitively on any non-trivial Steiner 5-design $\mathcal{D}$. Clearly, there exists always a Klein 4-group $V_4 \leq PSL(2, q)$, which fixes some 4-subset S of X and some additional point $x \in X$, and hence must fix the unique block $B \in \mathcal{B}$ which is incident with $S \cup \{x\}$ by the definition of Steiner 5-designs. Examining the list of possible subgroups of $PSL(2, q)$ with their orbits on the projective line (cf. Lemmas 8.3-8.11), it follows that we only have to consider the possibility when $PSL(2, q)_B$ is conjugate to $PSL(2, \bar{q})$ with $\bar{q}^m = q$, $m \geq 1$, and by Lemma 8.7, we conclude that $k = \bar{q} + 1$. Applying condition (B) gives then

$$(q - 2)(q - 3)\,|PSL(2, q)_{0B}| = \bar{q}(\bar{q} - 1)(\bar{q} - 2)(\bar{q} - 3)s \qquad (9.12)$$

$$\text{with } |PSL(2, q)_{0B}| = \bar{q}(\bar{q} - 1) \cdot \begin{cases} s, \text{ or} \\ 1. \end{cases}$$

Since $q = 2^{s^u}$, we can write $\bar{q} = 2^{s^w}$ for some integer $0 \leq w \leq u$, and $q = \bar{q}^m = \bar{q}^{s^{u-w}}$. As we may assume that $k = \bar{q} + 1 = 2^{s^w} + 1 > 5$, it follows in particular that $w \geq 1$, and hence $s < 2^{s^w} = \bar{q}$. Thus, using equation (9.12), we obtain

$$(\bar{q}^{s^{u-w}} - 2)(\bar{q}^{s^{u-w}} - 3) = (q - 2)(q - 3) \leq (\bar{q} - 2)(\bar{q} - 3)s < (\bar{q}^2 - 2s)(\bar{q} - 3).$$

But, as clearly $u - w \geq 1$ (otherwise, $k = q + 1$, a contradiction to Corollary 1.17), this yields a contradiction for every prime s.

Now, let $p = 3$. Then $\mathrm{Aut}(N) = P\Gamma L(2, q) = PGL(2, q) \rtimes \langle \tau_\alpha \rangle$, and $P\Gamma L(2, q)_{0,1,\infty} = \langle \tau_\alpha \rangle$ as $PGL(2, q)$ is sharply 3-transitive. Again, we define $G^* = G \cap (PSL(2, q) \rtimes \langle \tau_\alpha \rangle)$ and write $G^* = PSL(2, q) \rtimes (G^* \cap \langle \tau_\alpha \rangle)$ as in equation (9.4). We distinguish in the following the cases $G = G^*$ and $[G : G^*] = 2$.

First, let $G = G^*$. Then, we have $\mathrm{Aut}(N) = P\Sigma L(2,q)$. If we assume that $\langle \tau_\alpha \rangle \leq P\Sigma L(2,q)_{0B}$ for some appropriate, unique block $B \in \mathcal{B}$, then $G = P\Sigma L(2,q)$, and as clearly $PSL(2,q) \cap \langle \tau_\alpha \rangle = 1$, we can apply equation (9.5). Thus, $PSL(2,q)$ must also be flag-transitive, which has already been considered. Therefore, we may assume that $\langle \tau_\alpha \rangle \not\leq P\Sigma L(2,q)_{0B}$. Let s be a prime divisor of $e = |\langle \tau_\alpha \rangle|$. As already mentioned, the normal subgroup $H := (P\Sigma L(2,q)_{0,1,\infty})^s \leq \langle \tau_\alpha \rangle$ of index s has precisely $p^s + 1$ distinct fixed points, and hence we have $G \cap H \leq G_{0B}$ for some appropriate, unique block $B \in \mathcal{B}$ by the definition of Steiner 5-designs. It can then be deduced exactly as for $p = 2$ that $e = s^u$ for some $u \in \mathbb{N}$. As $\langle \tau_\alpha \rangle \not\leq P\Sigma L(2,q)_{0B}$, we may, by applying Dedekind's law, assume that

$$G_{0B} = PSL(2,q)_{0B} \rtimes (G \cap H).$$

Thus, by Remark 4.15, we obtain

$$2(q-2)(q-3)\,|PSL(2,q)_{0B}|\,|G \cap H| = (k-1)(k-2)(k-3)(k-4)\,|G \cap \langle \tau_\alpha \rangle|.$$

Using that $k = |0^{G_B}| = [G_B : G_{0B}]$, we have more precisely

(A^*) if $G = PSL(2,q) \rtimes (G \cap H)$:

$$2(q-2)(q-3)\,|PSL(2,q)_{0B}| = (k-1)(k-2)(k-3)(k-4)$$

$$\text{with } |PSL(2,q)_{0B}| = \frac{|PSL(2,q)_B|}{k}, \quad \text{or}$$

(B^*) if $G = P\Sigma L(2,q)$:

$$2(q-2)(q-3)\,|PSL(2,q)_{0B}| = (k-1)(k-2)(k-3)(k-4)s$$

$$\text{with } |PSL(2,q)_{0B}| = \frac{|PSL(2,q)_B|}{k} \cdot \begin{cases} s, & \text{if } G_B = PSL(2,q)_B \rtimes \langle \tau_\alpha \rangle, \\ 1, & \text{if } G_B = PSL(2,q)_B \rtimes H. \end{cases}$$

Considering condition (A^*), we may argue exactly as in the earlier case $N = G$. Thus, only condition (B^*) has to be examined, and we will show in the following that here $G \leq \mathrm{Aut}(\mathcal{D})$ cannot act flag-transitively on any non-trivial Steiner 5-design $\mathcal{D}$. In view of the subgroups of $PSL(2,q)$ with their orbits on the projective line (Lemmas 8.3–8.11), we have to examine the following possibilities:

(i) $PSL(2,q)_B$ is conjugate to a cyclic subgroup of order c with $c \mid \frac{q \pm 1}{2}$ of $PSL(2,q)$, and $k = c$.

(ii) $PSL(2,q)_B$ is conjugate to a dihedral subgroup of order $2c$ with $c \mid \frac{q \pm 1}{2}$ of $PSL(2,q)$, and $k = c$ or $2c$.

(iii) $PSL(2,q)_B$ is conjugate to an elementary Abelian subgroup of order $\bar{q} \mid q$ of $PSL(2,q)$, and $k = \bar{q}$.

(iv) $PSL(2, q)_B$ is conjugate to a semi-direct product of an elementary Abelian subgroup of order $\bar{q} \mid q$ with a cyclic subgroup of order c of $PSL(2, q)$ with $c \mid \bar{q} - 1$ and $c \mid q - 1$, and $k = \bar{q}$ or $c\bar{q}$.

(v) $PSL(2, q)_B$ is conjugate to $PSL(2, \bar{q})$ with $\bar{q}^m = q$, $m \geq 1$, and $k = \bar{q} + 1$, $\bar{q}(\bar{q} - 1)$ if m is even, or $k = (\bar{q} + 1)\bar{q}(\bar{q} - 1)/2$.

(vi) $PSL(2, q)_B$ is conjugate to $PGL(2, \bar{q})$ with $\bar{q}^m = q$, $m > 1$ even, and $k = \bar{q} + 1$, $\bar{q}(\bar{q} - 1)$ or $k = (\bar{q} + 1)\bar{q}(\bar{q} - 1)$.

(vii) $PSL(2, q)_B$ is conjugate to A_4, and $k = 6$ or 12.

(viii) $PSL(2, q)_B$ is conjugate to S_4, and $k = 6$ or 24.

(ix) $PSL(2, q)_B$ is conjugate to A_5, and $k = 10$, 12 or 60.

Since $q = 3^{s^u}$, we can write $\bar{q} = 3^{s^w}$ for some integer $0 \leq w \leq u$, and $q = \bar{q}^m = \bar{q}^{s^{u-w}}$.

ad (i): By condition (B*), we have

$$2(q - 2)(q - 3)\,|PSL(2, q)_{0B}| = (c - 1)(c - 2)(c - 3)(c - 4)s$$

$$\text{with } |PSL(2, q)_{0B}| = \begin{cases} s, \text{ or} \\ 1. \end{cases}$$

In view of the earlier case $N = G$, it is sufficient to consider the equation

$$(q - 2)(q - 3) = \frac{(c - 1)(c - 2)(c - 3)(c - 4)s}{2}. \tag{9.13}$$

For $c \mid \frac{q+1}{2}$, equation (9.13) gives

$$c \mid \frac{(q + 1)(q - 6)}{2} = \frac{(q - 2)(q - 3)}{2} - 6 = \frac{(c - 1)(c - 2)(c - 3)(c - 4)s}{4} - 6$$
$$= \frac{cs}{4}(c^3 - 10c^2 + 35c - 50) + 6s - 6,$$

and thus $c \mid 6s - 6$ must hold. If $c \mid \frac{q-1}{2}$, then, by equation (9.13), we have

$$c \mid \frac{(q - 1)(q - 4)}{2} = \frac{(q - 2)(q - 3)}{2} - 1 = \frac{(c - 1)(c - 2)(c - 3)(c - 4)s}{4} - 1$$
$$= \frac{cs}{4}(c^3 - 10c^2 + 35c - 50) + 6s - 1,$$

and hence $c \mid 6s - 1$ must hold. As clearly $c < 6s$ in both cases, it follows from equation (9.13) that in particular

$$(3^{s^u} - 2)(3^{s^u-1} - 1) < \frac{c^4 s}{6} < 6^3 \cdot s^5,$$

which implies that $s^u \leq 7$. As $c \mid 6s - 6$ respectively $c \mid 6s - 1$, this leaves only a very small number of possibilities for k to check, which can easily be ruled out by hand using equation (9.13).

ad (ii): Let $k = c$. Applying condition (B*) implies that

$$2(q - 2)(q - 3)\,|PSL(2, q)_{0B}| = (c - 1)(c - 2)(c - 3)(c - 4)s$$

$$\text{with } |PSL(2, q)_{0B}| = 2 \cdot \begin{cases} s, \text{ or} \\ 1. \end{cases}$$

First, let $k = c$. Due to the earlier case $N = G$, it is sufficient to consider the equation

$$(q - 2)(q - 3) = \frac{(c - 1)(c - 2)(c - 3)(c - 4)s}{4}. \tag{9.14}$$

If $c \mid \frac{q+1}{2}$, then, by equation (9.14), we have

$$c \mid \frac{(q + 1)(q - 6)}{2} = \frac{(q - 2)(q - 3)}{2} - 6 = \frac{(c - 1)(c - 2)(c - 3)(c - 4)s}{8} - 6$$

$$= \frac{cs}{8}(c^3 - 10c^2 + 35c - 50) + 3s - 6,$$

and hence $c \mid 3s - 6$ must hold. For $c \mid \frac{q-1}{2}$, it follows from equation (9.14) that

$$c \mid \frac{(q - 1)(q - 4)}{2} = \frac{(q - 2)(q - 3)}{2} - 1 = \frac{(c - 1)(c - 2)(c - 3)(c - 4)s}{8} - 1$$

$$= \frac{cs}{8}(c^3 - 10c^2 + 35c - 50) + 3s - 1,$$

and thus $c \mid 3s - 1$ must hold. Obviously, we have $c < 3s$ in both cases, and therefore equation (9.14) gives in particular

$$4(3^{s^u} - 2)(3^{s^u - 1} - 1) < \frac{c^4 s}{3} < 3^3 \cdot s^5,$$

which implies that $s^u \leq 5$. Due to the fact that $c \mid 3s - 6$ respectively $c \mid 3s - 1$, we have only a very small number of possibilities for k to check, which can easily be ruled out by hand using equation (9.14). Now, let $k = 2c$. Due to condition (B*), we have

$$2(q - 2)(q - 3)\,|PSL(2, q)_{0B}| = (2c - 1)(2c - 2)(2c - 3)(2c - 4)s$$

$$\text{with } |PSL(2, q)_{0B}| = \begin{cases} s, \text{ or} \\ 1. \end{cases}$$

Again, it suffices to consider the equation

$$\frac{(q - 2)(q - 3)}{2} = (2c - 1)(c - 1)(2c - 3)(c - 2)s. \tag{9.15}$$

For $c \mid \frac{q+1}{2}$, equation (9.15) gives

$$c \mid \frac{(q+1)(q-6)}{2} = \frac{(q-2)(q-3)}{2} - 6 = (2c-1)(c-1)(2c-3)(c-2)s - 6$$
$$= cs(4c^3 - 20c^2 + 35c - 25) + 6s - 6,$$

and thus $c \mid 6s - 6$ must hold. If $c \mid \frac{q-1}{2}$, then due to equation (9.15), we have

$$c \mid \frac{(q-1)(q-4)}{2} = \frac{(q-2)(q-3)}{2} - 1 = (2c-1)(c-1)(2c-3)(c-2)s - 1$$
$$= cs(4c^3 - 20c^2 + 35c - 25) + 6s - 1,$$

and hence $c \mid 6s - 1$ must hold. As clearly $c < 6s$ in both cases, we deduce from equation (9.15) that in particular

$$(3^{s^u} - 2)(3^{s^u-1} - 1) < \frac{(2c)^4 s}{6} < 2^4 \cdot 6^3 \cdot s^5,$$

and hence it follows that $s^u \le 7$. Since we have $c \mid 6s - 6$ respectively $c \mid 6s - 1$, this leaves only a very small number of possibilities for k to check, which can easily be ruled out by hand using equation (9.15).

ad (iii): In view of condition (B*), we have

$$2(q-2)(q-3)\,|PSL(2,q)_{0B}| = (\bar{q}-1)(\bar{q}-2)(\bar{q}-3)(\bar{q}-4)s$$

$$\text{with } |PSL(2,q)_{0B}| = \begin{cases} s, \text{ or} \\ 1. \end{cases}$$

It suffices to consider the equation

$$2(q-2)(q-3) = (\bar{q}-1)(\bar{q}-2)(\bar{q}-3)(\bar{q}-4)s. \qquad (9.16)$$

As we may assume that $k = \bar{q} = 3^{s^w} > 5$, we have in particular $w \ge 1$, and hence $s < 3^{s^w} = \bar{q}$. Thus, using equation (9.16), we obtain

$$(\bar{q}^{s^{u-w}} - 2)(\bar{q}^{s^{u-w}} - 3) = (q-2)(q-3) < \bar{q}^4 s < \bar{q}^5.$$

But, as clearly $u - w \ge 1$ (otherwise $k = q$, a contradiction to Corollary 1.17), this yields a contradiction for $s \ge 3$. If $s = 2$, then $(\bar{q}^{2^{u-w}} - 2)(\bar{q}^{2^{u-w}} - 3) < 2\bar{q}^4$ must hold, which cannot be true for $u - w > 1$. Thus, let $u - w = 1$. Hence, it follows from equation (9.16) that, in particular,

$$\bar{q} - 2 \mid (q-2)(q-3) = \bar{q}^4 - 5\bar{q}^2 + 6,$$

which is impossible since $(\bar{q}^4 - 5\bar{q}^2 + 6, \bar{q} - 2) = (2, \bar{q} - 2) = 1$.

ad (iv): Let $k = \bar{q}$. By condition (B*), we have

$$2(q - 2)(q - 3)\,|PSL(2, q)_{0B}| = (\bar{q} - 1)(\bar{q} - 2)(\bar{q} - 3)(\bar{q} - 4)s$$

$$\text{with } |PSL(2, q)_{0B}| = c \cdot \begin{cases} s, \text{ or} \\ 1. \end{cases}$$

As $c \mid \bar{q} - 1$, we may argue, mutatis mutandis, as in subcase (iii). For $k = c\bar{q}$, condition (B*) gives

$$2(q - 2)(q - 3)\,|PSL(2, q)_{0B}| = (c\bar{q} - 1)(c\bar{q} - 2)(c\bar{q} - 3)(c\bar{q} - 4)s$$

$$\text{with } |PSL(2, q)_{0B}| = \begin{cases} s, \text{ or} \\ 1. \end{cases}$$

We may consider only the equation

$$2(q - 2)(q - 3) = (c\bar{q} - 1)(c\bar{q} - 2)(c\bar{q} - 3)(c\bar{q} - 4)s. \tag{9.17}$$

Then, $(q - 2)(q - 3) = q^2 - 5q + 6$ must be divisible by $c\bar{q} - 3$. Polynomial division with remainder gives

$$
q^2 - 5q + 6 = \left(\sum_{i=1}^{m} 3^{i-1} \frac{q^2}{(c\bar{q})^i} + \sum_{j=1}^{\overline{m}} 3^{j-1} \frac{\left(\left(\frac{3}{c}\right)^m - 5\right)q}{(c\bar{q})^j} \right)(c\bar{q} - 3)
$$
$$
+ \left(\frac{3}{c}\right)^{\overline{m}} \frac{\left(\left(\frac{3}{c}\right)^m - 5\right)q}{\overline{q^m}} + 6
$$

for a suitable $\overline{m} \in \mathbb{N}$ (such that

$$
\deg\left(\left(\frac{3}{c}\right)^{\overline{m}} \frac{\left(\left(\frac{3}{c}\right)^m - 5\right)q}{\overline{q^m}} + 6 \right) < \deg\left(c\bar{q} - 3 \right)
$$

as is well-known). As $c \mid q - 1$, clearly c is not divisible by 3. Thus, the remainder can be rewritten as

$$
\frac{\left(\left(\frac{3}{c}\right)^m - 5\right)}{c^{\overline{m}}} \cdot 3^{s^u - \overline{m}(s^w - 1)} + 6,
$$

and hence in order for the remainder to vanish, necessarily $s^u - \overline{m}(s^w - 1) = 1$ must hold. But then, we obtain $3^m = (-2c^{\overline{m}} + 5)c^m$, a contradiction.

ad (v): Let $k = \bar{q} + 1$. In view of condition (B*), we have

$$2(q - 2)(q - 3)\,|PSL(2, q)_{0B}| = \bar{q}(\bar{q} - 1)(\bar{q} - 2)(\bar{q} - 3)s$$

$$\text{with } |PSL(2, q)_{0B}| = \frac{\bar{q}(\bar{q} - 1)}{2} \cdot \begin{cases} s, \text{ or} \\ 1. \end{cases}$$

Again, it suffices to consider the equation

$$(q - 2)(q - 3) = (\bar{q} - 2)(\bar{q} - 3)s. \tag{9.18}$$

As we may assume that $k = \bar{q} + 1 = 3^{s^w} + 1 > 5$, it follows in particular that $w \geq 1$, and hence $s < 3^{s^w} = \bar{q}$. Thus, using equation (9.18), we obtain

$$(\bar{q}^{s^{u-w}} - 2)(\bar{q}^{s^{u-w}} - 3) = (q-2)(q-3) = (\bar{q}-2)(\bar{q}-3)s < (\bar{q}^2 - 2s)(\bar{q}-3).$$

Since $u - w \geq 1$ (otherwise $k = q + 1$, yielding a contradiction to Corollary 1.17), this gives a contradiction for every prime s. If $m > 1$ even and $k = \bar{q}(\bar{q}-1)$, then, in view of condition (B*), we have

$$2(q-2)(q-3)\,|PSL(2,q)_{0B}| = (\bar{q}^2 - \bar{q} - 1)(\bar{q}^2 - \bar{q} - 2)(\bar{q}^2 - \bar{q} - 3)(\bar{q}^2 - \bar{q} - 4)s$$

$$\text{with } |PSL(2,q)_{0B}| = \frac{(\bar{q}+1)}{2} \cdot \left\{ \begin{array}{l} s, \text{ or} \\ 1. \end{array} \right.$$

We may consider only the equation

$$(q-2)(q-3)(\bar{q}+1) = (\bar{q}^2 - \bar{q} - 1)(\bar{q}^2 - \bar{q} - 2)(\bar{q}^2 - \bar{q} - 3)(\bar{q}^2 - \bar{q} - 4)s.$$

As obviously $(\bar{q}^2 - \bar{q} - 1, \bar{q} + 1) = 1$, it follows that $\bar{q}^2 - \bar{q} - 1 \mid (q-2)(q-3)$ must hold. But, for $m > 1$ even, polynomial division with remainder gives

$$q^2 - 5q + 6 = \left(\sum_{i=1}^{m-1} n_i \frac{q^2}{\bar{q}^{i+1}} + \sum_{j=1}^{m} (n_j \cdot n_m + n_{j-1}(n_{m-1} - 5))\frac{q}{\bar{q}^j} \right)(\bar{q}^2 - \bar{q} - 1)$$

$$+ (n_{m+1} \cdot n_m + n_m(n_{m-1} - 5))\bar{q} + n_m^2 + n_{m-1}(n_{m-1} - 5) + 6,$$

where n_i denote the i-th Fibonacci number recursively defined via

$$n_0 = 0, \ n_1 = n_2 = 1, \ n_i = n_{i-1} + n_{i-2} \ (i \geq 3).$$

As can easily be seen, the remainder never vanishes, and hence we obtain a contradiction. For $k = (\bar{q}+1)\bar{q}(\bar{q}-1)/2$, condition (B*) gives

$$2(q-2)(q-3)\,|PSL(2,q)_{0B}| = (\frac{\bar{q}^3 - \bar{q}}{2} - 1)(\frac{\bar{q}^3 - \bar{q}}{2} - 2)(\frac{\bar{q}^3 - \bar{q}}{2} - 3)(\frac{\bar{q}^3 - \bar{q}}{2} - 4)s$$

$$\text{with } |PSL(2,q)_{0B}| = \left\{ \begin{array}{l} s, \text{ or} \\ 1. \end{array} \right.$$

It suffices to consider the equation

$$2(q-2)(q-3) = (\frac{\bar{q}^3 - \bar{q}}{2} - 1)(\frac{\bar{q}^3 - \bar{q}}{2} - 2)(\frac{\bar{q}^3 - \bar{q}}{2} - 3)(\frac{\bar{q}^3 - \bar{q}}{2} - 4)s. \quad (9.19)$$

If we assume that $\bar{q} = 3$, then $k = 12$. Thus, we obtain from equation (9.19) that $s^u < 5$. Hence, there are only a very small number of possibilities to check, which can easily be ruled out by hand. Therefore, let us assume that $\bar{q} > 3$. Then, we have in particular $w \geq 1$, and hence $s < 3^{s^w} = \bar{q}$. Thus, using equation (9.19), we obtain

$$2(q-2)(q-3) < (\frac{\bar{q}^3 - \bar{q}}{2})^4 s < \frac{1}{16}\bar{q}^{12}s < \frac{1}{16}\bar{q}^{13}.$$

On the other hand, it follows that

$$
\begin{aligned}
2(q-2)(q-3) &= (\frac{\bar{q}^3 - \bar{q}}{2} - 1)(\frac{\bar{q}^3 - \bar{q}}{2} - 2)(\frac{\bar{q}^3 - \bar{q}}{2} - 3)(\frac{\bar{q}^3 - \bar{q}}{2} - 4)s \\
&\geq 2(\frac{\bar{q}^3 - \bar{q}}{2} - 1)(\frac{\bar{q}^3 - \bar{q}}{2} - 2)(\frac{\bar{q}^3 - \bar{q}}{2} - 3)(\frac{\bar{q}^3 - \bar{q}}{2} - 4) \\
&= \frac{1}{8}\bar{q}^{12} - l
\end{aligned}
$$

with $l = \frac{1}{2}\bar{q}^{10} + \frac{5}{2}\bar{q}^9 - \frac{3}{4}\bar{q}^8 - \frac{15}{2}\bar{q}^7 - 17\bar{q}^6 + \frac{15}{2}\bar{q}^5 + \frac{279}{8}\bar{q}^4 + \frac{95}{2}\bar{q}^3 - \frac{35}{2}\bar{q}^2 - 50\bar{q} - 48$.
As for $\bar{q} > 3$, clearly $l < \frac{1}{16}\bar{q}^{12}$ holds, thus we obtain

$$2(q-2)(q-3) \geq \frac{1}{16}\bar{q}^{12}.$$

But as $2(q-2)(q-3) = 2(\bar{q}^{2m} - 5\bar{q}^m + 6)$, this leaves at most only $m = 6$, which clearly cannot occur since $m = s^{u-w}$.

ad (vi): Let $k = \bar{q} + 1$. Applying condition (B*) gives

$$2(q-2)(q-3)|PSL(2,q)_{0B}| = \bar{q}(\bar{q}-1)(\bar{q}-2)(\bar{q}-3)s$$

$$\text{with } |PSL(2,q)_{0B}| = \bar{q}(\bar{q}-1) \cdot \begin{cases} s, \text{ or} \\ 1. \end{cases}$$

Clearly, we may argue, mutatis mutandis, as for $k = \bar{q} + 1$ in subcase (v). Let $k = \bar{q}(\bar{q} - 1)$. Due to condition (B*), we have

$$2(q-2)(q-3)|PSL(2,q)_{0B}| = (\bar{q}^2 - \bar{q} - 1)(\bar{q}^2 - \bar{q} - 2)(\bar{q}^2 - \bar{q} - 3)(\bar{q}^2 - \bar{q} - 4)s$$

$$\text{with } |PSL(2,q)_{0B}| = (\bar{q}+1) \cdot \begin{cases} s, \text{ or} \\ 1. \end{cases}$$

As $\bar{q}^2 - \bar{q} - 1$ is always odd, we may argue, mutatis mutandis, as for $k = \bar{q}(\bar{q} - 1)$ in subcase (v). Now, let $k = (\bar{q} + 1)\bar{q}(\bar{q} - 1)$. Then, condition (B*) implies that

$$2(q-2)(q-3)|PSL(2,q)_{0B}| = (\bar{q}^3 - \bar{q} - 1)(\bar{q}^3 - \bar{q} - 2)(\bar{q}^3 - \bar{q} - 3)(\bar{q}^3 - \bar{q} - 4)s$$

$$\text{with } |PSL(2,q)_{0B}| = \begin{cases} s, \text{ or} \\ 1. \end{cases}$$

We may consider only the equation

$$2(q-2)(q-3) = (\bar{q}^3 - \bar{q} - 1)(\bar{q}^3 - \bar{q} - 2)(\bar{q}^3 - \bar{q} - 3)(\bar{q}^3 - \bar{q} - 4)s. \quad (9.20)$$

If $\bar{q} = 3$, then we have $k = 24$, which implies that $s^u \leq 5$ must hold. Hence, we have only a very small number of possibilities to check, which can easily be ruled out by hand. Thus, let us assume that $\bar{q} > 3$. Then, it follows in particular that $w \geq 1$, and hence $s < 3^{s^w} = \bar{q}$. Using equation (9.20), it follows therefore that

$$2(q-2)(q-3) < (\bar{q}^3 - \bar{q})^4 s < \bar{q}^{12} s < \bar{q}^{13}.$$

On the other hand, we have

$$
\begin{aligned}
2(q-2)(q-3) &= (\bar{q}^3 - \bar{q} - 1)(\bar{q}^3 - \bar{q} - 2)(\bar{q}^3 - \bar{q} - 3)(\bar{q}^3 - \bar{q} - 4)s \\
&\geq 2(\bar{q}^3 - \bar{q} - 1)(\bar{q}^3 - \bar{q} - 2)(\bar{q}^3 - \bar{q} - 3)(\bar{q}^3 - \bar{q} - 4) \\
&= 2\bar{q}^{12} - l
\end{aligned}
$$

with $l = 8\bar{q}^{10} + 20\bar{q}^9 - 12\bar{q}^8 - 60\bar{q}^7 - 62\bar{q}^6 + 60\bar{q}^5 + 138\bar{q}^4 + 80\bar{q}^3 - 70\bar{q}^2 - 100\bar{q} - 48$. As for $\bar{q} > 3$, clearly $l < \bar{q}^{12}$ holds, it follows that

$$2(q-2)(q-3) \geq \bar{q}^{12}.$$

But as $2(q-2)(q-3) = 2(\bar{q}^{2m} - 5\bar{q}^m + 6)$, this leaves only $m = 6$, which obviously cannot occur since $m = s^{u-w}$.

ad (vii): In view of condition (B*), we have

$$2(q-2)(q-3)\,|PSL(2,q)_{0B}| = (k-1)(k-2)(k-3)(k-4)s$$

$$\text{with } |PSL(2,q)_{0B}| = \frac{12}{k} \cdot \begin{cases} s, \text{ or} \\ 1. \end{cases}$$

It is sufficient to consider the equation

$$(3^{s^u} - 2)(3^{s^u} - 3) = \frac{k(k-1)(k-2)(k-3)(k-4)}{24} \cdot s.$$

Thus, for $k = 6$, respectively $k = 12$, we obtain $s^u \leq 2$, respectively $s^u < 5$, and thus we have only a very small number of possibilities to check, which can easily be ruled out by hand.

ad (viii): Applying condition (B*) gives

$$2(q-2)(q-3)\,|PSL(2,q)_{0B}| = (k-1)(k-2)(k-3)(k-4)s$$

$$\text{with } |PSL(2,q)_{0B}| = \frac{24}{k} \cdot \begin{cases} s, \text{ or} \\ 1. \end{cases}$$

We may consider only the equation

$$2(3^{s^u} - 2)(3^{s^u} - 3) = \frac{k(k-1)(k-2)(k-3)(k-4)}{24} \cdot s.$$

Thus, for $k = 6$, respectively $k = 24$, we obtain $s^u < 2$, respectively $s^u \leq 5$, and hence we have only a very small number of possibilities to check, which can again easily be ruled out by hand.

ad (ix): Due to condition (B*), we have

$$2(q-2)(q-3)\,|PSL(2,q)_{0B}| = (k-1)(k-2)(k-3)(k-4)s$$

$$\text{with } |PSL(2,q)_{0B}| = \frac{60}{k} \cdot \begin{cases} s, \text{ or} \\ 1. \end{cases}$$

It is sufficient to consider the equation

$$5(3^{s^u}-2)(3^{s^u}-3) = \frac{k(k-1)(k-2)(k-3)(k-4)}{24} \cdot s.$$

Thus, for $k = 10$ and 12, it follows in both cases that $s^u \leq 3$, and for $k = 60$, we obtain $s^u \leq 7$. Hence, we have again only a very small number of possibilities to check, which can easily be ruled out by hand again, completing the examination of condition (B*).

Now, we consider the case when $[G : G^*] = 2$. We first assume that $\langle \tau_\alpha \rangle \not\leq P\Gamma L(2,q)_{0B}$ for some appropriate, unique block $B \in \mathcal{B}$. Again, let s be a prime divisor of $e = |\langle \tau_\alpha \rangle|$. As the normal subgroup $H := (P\Gamma L(2,q)_{0,1,\infty})^s \leq \langle \tau_\alpha \rangle$ of index s has precisely $p^s + 1$ distinct fixed points, we have $G \cap H \leq G_{0B}$ for some appropriate, unique block $B \in \mathcal{B}$ by the definition of Steiner 5-designs. Mutatis mutandis as in case $p = 2$, it follows then that $e = s^u$ for some $u \in \mathbb{N}$. Since $\langle \tau_\alpha \rangle \not\leq P\Gamma L(2,q)_{0B}$, we may, by applying Dedekind's law, assume that

$$G_{0B} = \begin{cases} PGL(2,q)_{0B} \rtimes (G \cap H), \text{ or} \\ PSL(2,q)_{0B} \rtimes (G \cap H). \end{cases}$$

Thus, by Remark 4.15, we obtain

$$(q-2)(q-3) \cdot \begin{cases} |PGL(2,q)_{0B}| \\ |PSL(2,q)_{0B}| \end{cases} \cdot |G \cap H| = (k-1)(k-2)(k-3)(k-4)\,|G \cap \langle \tau_\alpha \rangle|.$$

Using that $k = |0^{G_B}| = [G_B : G_{0B}]$, we have more precisely

(Ā) if $G = PGL(2,q) \rtimes (G \cap H)$:

(i) for $G_{0B} = PGL(2,q)_{0B} \rtimes (G \cap H)$:

$$2(q-2)(q-3)\,|PSL(2,q)_{0B}| = (k-1)(k-2)(k-3)(k-4)$$

$$\text{with } |PSL(2,q)_{0B}| = \frac{|PSL(2,q)_B|}{k}, \text{ or}$$

(ii) for $G_{0B} = PSL(2,q)_{0B} \rtimes (G \cap H)$:

$$(q-2)(q-3)\,|PSL(2,q)_{0B}| = (k-1)(k-2)(k-3)(k-4)$$

with

$$|PSL(2,q)_{0B}| = \frac{|PSL(2,q)_B|}{k} \cdot \begin{cases} 2, \text{ if } G_B = PGL(2,q)_B \rtimes (G \cap H) \\ 1, \text{ if } G_B = PSL(2,q)_B \rtimes (G \cap H), \end{cases}$$

or

($\bar{\mathrm{B}}$) if $G = P\Gamma L(2, q)$:

(i) for $G_{0B} = PGL(2, q)_{0B} \rtimes H$:

$$2(q-2)(q-3)\,|PSL(2,q)_{0B}| = (k-1)(k-2)(k-3)(k-4)s$$

with $|PSL(2,q)_{0B}| = \dfrac{|PSL(2,q)_B|}{k} \cdot \begin{cases} s, & \text{if } G_B = PGL(2,q)_B \rtimes \langle \tau_\alpha \rangle \\ 1, & \text{if } G_B = PGL(2,q)_B \rtimes H, \text{ or} \end{cases}$

(ii) for $G_{0B} = PSL(2, q)_{0B} \rtimes H$:

$$(q-2)(q-3)\,|PSL(2,q)_{0B}| = (k-1)(k-2)(k-3)(k-4)s$$

with $|PSL(2,q)_{0B}| = \dfrac{|PSL(2,q)_B|}{k} \cdot \begin{cases} 2s, & \text{if } G_B = PGL(2,q)_B \rtimes \langle \tau_\alpha \rangle, \\ s, & \text{if } G_B = PSL(2,q)_B \rtimes \langle \tau_\alpha \rangle, \\ 2, & \text{if } G_B = PGL(2,q)_B \rtimes H, \\ 1, & \text{if } G_B = PSL(2,q)_B \rtimes H. \end{cases}$

As far as condition ($\bar{\mathrm{A}}$) is concerned, we may argue exactly as in the earlier case $N = G$. Therefore, only condition ($\bar{\mathrm{B}}$) has to be examined, and we may argue, mutatis mutandis, as for condition (B*) to prove that here $G \leq \mathrm{Aut}(\mathcal{D})$ cannot act flag-transitively on any non-trivial Steiner 5-design $\mathcal{D}$. Finally, let us assume that $\langle \tau_\alpha \rangle \leq P\Gamma L(2,q)_{0B}$. Then, $G = P\Gamma L(2,q)$ and, by Dedekind's law, we can write

$$G_{0B} = \begin{cases} PGL(2,q)_{0B} \rtimes \langle \tau_\alpha \rangle, \text{ or} \\ PSL(2,q)_{0B} \rtimes \langle \tau_\alpha \rangle. \end{cases}$$

Hence, Remark 4.15 yields

$$(q-2)(q-3) \cdot \begin{cases} |PGL(2,q)_{0B}| \\ |PSL(2,q)_{0B}| \end{cases} \cdot |\langle \tau_\alpha \rangle| = (k-1)(k-2)(k-3)(k-4)\,|\langle \tau_\alpha \rangle|.$$

As $k = |0^{G_B}| = [G_B : G_{0B}]$, we obtain more precisely

(i) for $G_{0B} = PGL(2,q)_{0B} \rtimes \langle \tau_\alpha \rangle$:

$$2(q-2)(q-3)\,|PSL(2,q)_{0B}| = (k-1)(k-2)(k-3)(k-4)$$

with $|PSL(2,q)_{0B}| = \dfrac{|PSL(2,q)_B|}{k}, \text{ or}$

(ii) for $G_{0B} = PSL(2,q)_{0B} \rtimes \langle \tau_\alpha \rangle$:

$$(q-2)(q-3)\,|PSL(2,q)_{0B}| = (k-1)(k-2)(k-3)(k-4)$$

with $|PSL(2,q)_{0B}| = \dfrac{|PSL(2,q)_B|}{k} \cdot \begin{cases} 2, & \text{if } G_B = PGL(2,q)_B \rtimes \langle \tau_\alpha \rangle, \\ 1, & \text{if } G_B = PSL(2,q)_B \rtimes \langle \tau_\alpha \rangle. \end{cases}$

Again, we may argue here exactly as in the earlier case $N = G$, and the claim follows.

Case (3): $N = M_v$, $v = 11, 12, 22, 23, 24$.

If $v = 12$ or 24, then $G = M_v$ is always 5-transitive, and thus Theorem 5.2 gives the designs described in Theorem 9.1. Obviously, flag-transitivity holds as the 5-transitivity of G implies that G_x acts block-transitively on the derived Steiner 4-design $\mathcal{D}_x$ for any $x \in X$. By Corollary 1.17, we obtain for $v = 11$ that $k \leq 6$, and for $v = 22$ or 23 that $k \leq 8$, and the very small number of cases for k can easily be ruled out by hand using Lemma 4.1.

Case (4): $N = M_{11}$, $v = 12$.

By the same arguments as in the corresponding case in Theorem 8.1, it follows that $G \leq \mathrm{Aut}(\mathcal{D})$ cannot act on any Steiner t-design $\mathcal{D}$.

This completes the proof of Theorem 9.1. □

Chapter 10

The Non-Existence of Flag-transitive Steiner 6-Designs

10.1 Introduction

Relying on the classification of the finite 3-homogeneous permutation groups, we easily prove in this chapter that there are no non-trivial flag-transitive Steiner 6-designs. In addition, we state in this context an interesting conjecture of P. J. Cameron and C. E. Praeger [29, Conj. 1.2] on block-transitive 6-designs.

10.2 Main Result

We state our result:

Theorem 10.1. *There exist no non-trivial Steiner 6-design $\mathcal{D}$ admitting a flag-transitive group $G \leq \mathrm{Aut}(\mathcal{D})$ of automorphisms.*

10.3 Groups of Automorphisms of Affine Type

In the following, we begin with the proof of Theorem 10.1. Let us assume that $\mathcal{D} = (X, \mathcal{B}, I)$ is a non-trivial Steiner 6-design with $G \leq \mathrm{Aut}(\mathcal{D})$ acting flag-transitively on $\mathcal{D}$ throughout this chapter. We recall that due to Proposition 4.14, we may restrict ourselves to the inspection of the finite 3-homogeneous permutation groups. Clearly, in the sequel we may assume that $k > 6$ as trivial Steiner 6-designs are excluded. We will examine in this section successively those cases where G is of affine type.

Case (1): $G \cong AGL(1,8)$, $A\Gamma L(1,8)$ or $A\Gamma L(1,32)$.

We may assume that $k > 6$. If $v = 8$, then Corollary 1.17 would imply that $k = 6$. For $v = 32$, we have $|G| = 5v(v-1)$ and Lemma 4.1 immediately implies that $G \leq \mathrm{Aut}(\mathcal{D})$ cannot act flag-transitively on any non-trivial Steiner 6-design $\mathcal{D}$.

Case (2): $G_0 \cong SL(d,2)$, $d \geq 2$.

We may assume that $v = 2^d > k > 6$. For $d = 3$, we have $v = 8$ and $k = 7$ by Corollary 1.17, which is not possible in view of Lemma 1.14 (c). For $d > 3$, we may argue mutatis mutandis, as in the corresponding case in Theorem 9.1, to obtain that each block must be contained in a 3-dimensional affine subspace, and hence $k \leq 8$. On the other hand, for $\mathcal{D}$ to be a block-transitive 6-design admitting $G \leq \mathrm{Aut}(\mathcal{D})$, we deduce from [29, Prop. 3.6 (b)] the necessary condition that $2^d - 3$ must divide $\binom{k}{4}$, which implies for each respective value of k that $d = 3$ must hold, a contradiction.

Case (3): $G_0 \cong A_7$, $v = 2^4$.

As $v = 2^4$, we have $k \leq 8$ by Corollary 1.17. But, Lemma 1.14 (c) obviously eliminates the cases when $k = 7$ or 8.

10.4   Groups of Automorphisms of Almost Simple Type

We will examine in this section successively those cases where G is of almost simple type.

Case (1): $N = A_v$, $v \geq 5$.

We may assume that $v \geq 8$. But then A_v, and hence also G, is 6-transitive and does not act on any non-trivial Steiner 6-design $\mathcal{D}$ due to Theorem 5.2.

Case (2): $N = PSL(2,q)$, $v = q+1$, $q = p^e > 3$.

For the existence of flag-transitive Steiner 6-designs, necessarily

$$r = \frac{q(q-1)(q-2)(q-3)(q-4)}{(k-1)(k-2)(k-3)(k-4)(k-5)} \,\Big|\, |G_0| \,\Big|\, |P\Gamma L(2,q)_0| = q(q-1)e$$

must hold in view of Lemma 4.1. Thus, we have in particular

$$(q-2)(q-3)(q-4) \mid (k-1)(k-2)(k-3)(k-4)(k-5)e, \text{ where } e \leq \log_2 q. \quad (10.1)$$

On the other hand, Corollary 1.17 implies $k \leq \lfloor \sqrt{q+1} + \frac{9}{2} \rfloor < q^{\frac{1}{2}} + 5$. Hence, in view of property (10.1), we have only a small number of possibilities to check, which can easily be ruled out by hand using Lemma 1.14 (c). Therefore, $G \leq \mathrm{Aut}(\mathcal{D})$ cannot act flag-transitively on any non-trivial Steiner 6-design $\mathcal{D}$. This has also been proven in [29, Cor. 4.3], whereas our estimation is slightly better.

Case (3): $N = M_v$, $v = 11, 12, 22, 23, 24$.

Due to Corollary 1.17, we obtain for $v = 11$ or 12 that $k \leq 7$, and for $v = 22, 23$ or 24 that $k \leq 9$, and the very small number of cases for k can easily be eliminated by hand using Lemma 4.1.

Case (4): $N = M_{11}$, $v = 12$.

By the same arguments as in the corresponding case in Theorem 8.1, it follows that $G \leq \mathrm{Aut}(\mathcal{D})$ cannot act on any Steiner t-design $\mathcal{D}$.

This completes the proof of Theorem 10.1. □

In closing this chapter, we state in this context a far-reaching conjecture of P. J. Cameron and C. E. Praeger [29, Conj. 1.2] on block-transitive 6-designs:

Conjecture 10.2. (Cameron and Praeger 1993). There exist no non-trivial 6-design $\mathcal{D}$ admitting a block-transitive group $G \leq \mathrm{Aut}(\mathcal{D})$ of automorphisms.

Remark 10.3. We note that very recently an important step has been taken towards confirming this conjecture: We essentially proved that the Cameron-Praeger conjecture is true for the important case of Steiner 6-designs. The result is announced in [64], and the longer proof will appear shortly [66]. Hence, with regard to highly symmetric Steiner designs, this result together with Theorem 10.1 contributes to the fundamental problem in design theory whether there exists any non-trivial Steiner 6-design (cf. Problem 1.12).

Bibliography

[1] W. O. Alltop, *5-designs in affine spaces*, Pacific J. Math. **39** (1971), 547–551.

[2] M. Aschbacher, *Chevalley groups of type G_2 as the group of a trilinear form*, J. Algebra **109** (1987), 193–259.

[3] ———, *Sporadic Groups*, Cambridge Tracts in Math. **104**, Cambridge Univ. Press, Cambridge, 1994.

[4] ———, *Finite Group Theory*, Cambridge Studies in Advanced Math. **10**, 2nd ed., Cambridge Univ. Press, Cambridge, 2000.

[5] E. F. Assmus, Jr. and J. D. Key, *Designs and their Codes*, Cambridge Tracts in Math. **103**, Cambridge Univ. Press, Cambridge, 1993.

[6] R. Baer, *Polarities in finite projective planes*, Bull. Amer. Math. Soc. **52** (1946), 77–93.

[7] J. A. Barrau, *On the Combinatory Problem of Steiner*, Proc. Sect. Sci. Konink. Akad. Wetensch. Amsterdam **11** (1908), 352–360.

[8] Th. Beth, D. Jungnickel, and H. Lenz, *Design Theory*, Vol. I and II, Encyclopedia of Math. and Its Applications **69/78**, Cambridge Univ. Press, Cambridge, 1999.

[9] A. Betten, R. Laue, and A. Wassermann, *A Steiner 5-design on 36 points*, Designs, Codes, Cryptography **17** (1999), 181–186.

[10] F. Beukers, *On the generalized Ramanujan-Nagell equation, I*, Acta Arith. **38** (1980/81), 389–410.

[11] ———, *On the generalized Ramanujan-Nagell equation, II*, Acta Arith. **39** (1981), 113–123.

[12] A. Beutelspacher, *Einführung in die endliche Geometrie I: Blockpläne*, Bibliographisches Institut, Mannheim, Wien, Zürich, 1982.

[13] ———, *Projective planes*, in: Handbook of Incidence Geometry, ed. by F. Buekenhout, North-Holland, Amsterdam, New York, Oxford, 1995, 107–136.

[14] G. D. Birkhoff and H. S. Vandiver, *On the integral divisors of $a^n - b^n$*, Ann. Math. **5** (1904), 173–180.

[15] R. E. Block, *Transitive groups of collineations on certain designs*, Pacific J. Math. **15** (1965), 13–18.

[16] _____, *On the orbits of collineation groups*, Math. Z. **96** (1967), 33–49.

[17] R. C. Bose and S. S. Shrikhande, *On the construction of sets of mutually orthogonal Latin squares and the falsity of a conjecture of Euler*, Trans. Amer. Math. Soc. **95** (1960), 191–209.

[18] R. H. Bruck and H. J. Ryser, *The non-existence of certain finite projective planes*, Canad. J. Math. **1** (1949), 88–93.

[19] F. Buekenhout, *Remarques sur l'homogénéité des espaces linéaires et des systèmes de blocs*, Math. Z. **104** (1968), 144–146.

[20] _____, *Flag-transitive Steiner systems after Michael Huber*, unpublished manuscript (2006).

[21] F. Buekenhout, P.-O. Dehaye, and D. Leemans, *RWPRI and $(2T)_1$ flag-transitive linear spaces*, Contr. Alg. Geom. **44** (2003), 25–46.

[22] F. Buekenhout, A. Delandtsheer, and J. Doyen, *Finite linear spaces with flag-transitive groups*, J. Combin. Theory, Series A **49** (1988), 268–293.

[23] F. Buekenhout, A. Delandtsheer, J. Doyen, P. B. Kleidman, M. W. Liebeck, and J. Saxl, *Linear spaces with flag-transitive automorphism groups*, Geom. Dedicata **36** (1990), 89–94.

[24] P. J. Cameron, *Parallelisms of Complete Designs*, London Math. Soc. Lecture Note Series **23**, Cambridge Univ. Press, Cambridge, 1976.

[25] _____, *Finite permutation groups and finite simple groups*, Bull. London Math. Soc. **13** (1981), 1–22.

[26] _____, *Permutation Groups*, London Math. Soc. Student Texts **45**, Cambridge Univ. Press, Cambridge, 1999.

[27] P. J. Cameron and W. M. Kantor, *2-transitive and antiflag transitive collineation groups of finite projective and polar spaces*, J. Algebra **60** (1979), 384–422.

[28] P. J. Cameron, H. R. Maimani, G. R. Omidi, and B. Tayfeh-Rezaie, *3-designs from $PSL(2, q)$*, Discrete Math. **306** (2006), 3063–3073.

[29] P. J. Cameron and C. E. Praeger, *Block-transitive t-designs, II: large t*, in: Finite Geometry and Combinatorics (Deinze 1992), ed. by F. De Clerck et al., London Math. Soc. Lecture Note Series **191**, Cambridge Univ. Press, Cambridge, 1993, 103–119.

[30] P. J. Cameron and J. H. van Lint, *Designs, Graphs, Codes and their Links*, London Math. Soc. Student Texts **22**, Cambridge Univ. Press, Cambridge, 1991.

[31] R. D. Carmichael, *Introduction to the Theory of Groups of Finite Order*, Ginn, Boston, 1937; Reprint: Dover Publications, New York, 1956.

[32] R. W. Carter, *Simple Groups of Lie Type*, J. Wiley, New York, 1972; Reprint: J. Wiley, 1989.

[33] P. C. Clapham, *Steiner triple systems with block-transitive automorphism groups*, Discrete Math. **14** (1976), 121–131.

[34] C. J. Colbourn and J. H. Dinitz (Eds.), *Handbook of Combinatorial Designs*, 2nd ed., Discrete Math. and Its Applications **42**, CRC Press, Boca Raton, 2006.

[35] J. H. Conway, R. T. Curtis, S. P. Norton, R. A. Parker, and R. A. Wilson, *Atlas of Finite Groups*, Clarendon Press, Oxford, 1985.

[36] J. H. Conway and N. J. A. Sloane, *Sphere Packings, Lattices and Groups*, 3rd ed., Springer, Berlin, Heidelberg, New York, 1998.

[37] C. W. Curtis, W. M. Kantor, and G. M. Seitz, *The 2-transitive permutation representations of the finite Chevalley groups*, Trans. Amer. Math. Soc. **218** (1976), 1–59.

[38] A. Delandtsheer, *Finite (line, plane)-flag-transitive planar spaces*, Geom. Dedicata **41** (1992), 145–153.

[39] _____, *Dimensional linear spaces*, in: Handbook of Incidence Geometry, ed. by F. Buekenhout, North-Holland, Amsterdam, New York, Oxford, 1995, 193–294.

[40] _____, *Finite flag-transitive linear spaces with alternating socle*, in: Algebraic Combinatorics and Applications, Proc. Euroconf. (Gößweinstein 1999), ed. by A. Betten et al., Springer, Berlin, 2001, 79–88.

[41] A. Delandtsheer, J. Doyen, J. Siemons, and C. Tamburini, *Doubly homogeneous 2-(v, k, 1) designs*, J. Combin. Theory, Series A **43** (1986), 140–145.

[42] P. Dembowski, *Verallgemeinerungen von Transitivitätsklassen endlicher projektiver Ebenen*, Math. Z. **69** (1958), 59–89.

[43] _____, *Finite Geometries*, Classics in Math., Springer, Berlin, Heidelberg, New York, 1968; Reprint 1997.

[44] L. E. Dickson, *Linear Groups with an Exposition of the Galois Field Theory*, Teubner, Leipzig, 1901; Reprint: Dover Publications, New York, 1958.

[45] J. D. Dixon and B. Mortimer, *Permutation Groups*, Graduate Texts in Math. **163**, Springer, Berlin, Heidelberg, New York, 1996.

[46] F. Buekenhout (Ed.), *Handbook of Incidence Geometry*, North-Holland, Amsterdam, New York, Oxford, 1995.

[47] R. A. Fisher, *An examination of the different possible solutions of a problem in incomplete blocks*, Ann. Eugenics **10** (1940), 52–75.

[48] D. A. Foulser, *The flag-transitive collineation groups of the finite desarguesian affine planes*, Canad. J. Math. **16** (1964), 443–472.

[49] _____, *Solvable flag-transitive affine groups*, Math. Z. **86** (1964), 191–204.

[50] D. Gorenstein, *Finite Simple Groups. An Introduction to Their Classification*, Plenum Publishing Corp., New York, London, 1982.

[51] M. Hall, Jr., *Combinatorial Theory*, 2nd ed., J. Wiley, New York, 1986.

[52] H. Hanani, *On quadruple systems*, Canad. J. Math. **12** (1960), 145–157.

[53] H. Hasse, *Über eine diophantische Gleichung von Ramanujan-Nagell und ihre Verallgemeinerung*, Nagoya Math. J. **27** (1966), 77–102.

[54] C. Hering, *Transitive linear groups and linear groups which contain irreducible subgroups of prime order*, Geom. Dedicata **2** (1974), 425–460.

[55] _____, *Transitive linear groups and linear groups which contain irreducible subgroups of prime order, II*, J. Algebra **93** (1985), 151–164.

[56] D. G. Higman and J. E. McLaughlin, *Geometric ABA-groups*, Illinois J. Math. **5** (1961), 382–397.

[57] J. W. P. Hirschfeld, *Projective Geometries over Finite Fields*, Oxford Math. Monographs, 2nd ed., Clarendon Press, Oxford, 1998.

[58] J. Höchsmann, *On minimal p-degrees in 2-transitive permutation groups*, Arch. Math. **72** (1999), 405–417.

[59] M. Huber, *Classification of flag-transitive Steiner quadruple systems*, J. Combin. Theory, Series A **94** (2001), 180–190.

[60] _____, *The classification of flag-transitive Steiner 3-designs*, Adv. Geom. **5** (2005), 195–221.

[61] _____, *On Highly Symmetric Combinatorial Designs*, Habilitationsschrift, Univ. Tübingen, Tübingen, 2005, Shaker, Aachen, 2006.

[62] _____, *A census of highly symmetric combinatorial designs*, J. Algebr. Comb. **26** (2007), 453–476.

[63] _____, *The classification of flag-transitive Steiner 4-designs*, J. Algebr. Comb. **26** (2007), 183–207.

[64] _____, *Steiner t-designs for large t*, in: Math. Methods in Comp. Science (MMICS 2008), ed. by J. Calmet, W. Geiselmann and J. Müller-Quade, Lecture Notes in Comp. Science 5393 (Beth Festschrift), Springer, Berlin, Heidelberg, New York, 2008, 18–26.

[65] _____, *Coding theory and algebraic combinatorics*, in: Selected Topics in Information and Coding Theory, ed. by I. Woungang et al., World Scientific, Singapore, 33 pages (to appear).

[66] _____, *On the Cameron-Praeger conjecture*, (submitted).

[67] W. C. Huffman and V. Pless (Eds.), *Handbook of Coding Theory*, Vol. I and II, North-Holland, Amsterdam, New York, Oxford, 1998.

[68] W. C. Huffman and V. Pless, *Fundamentals of Error-Correcting Codes*, Cambridge Univ. Press, Cambridge, 2003.

[69] D. R. Hughes and F. C. Piper, *Projective Planes*, Graduate Texts in Math. **6**, 2nd ed., Springer, Berlin, Heidelberg, New York, 1982.

[70] _____, *Design Theory*, Cambridge Univ. Press, Cambridge, 1985.

[71] S. H. Y. Hung and N. S. Mendelsohn, *On the Steiner systems $S(3,4,14)$ and $S(4,5,15)$*, Utilitas Math. **1** (1972), 5–95.

[72] B. Huppert, *Zweifach transitive, auflösbare Permutationsgruppen*, Math. Z. **68** (1957), 126–150.

[73] _____, *Endliche Gruppen I*, Grundlehren der Math. Wissenschaften **134**, Springer, Berlin, Heidelberg, New York, 1967.

[74] B. Huppert and N. Blackburn, *Finite Groups III*, Grundlehren der Math. Wissenschaften **243**, Springer, Berlin, Heidelberg, New York, 1982.

[75] Y. J. Ionin and M. S. Shrikhande, *Combinatorics of Symmetric Designs*, New Math. Monographs **5**, Cambridge Univ. Press, Cambridge, 2006.

[76] W. M. Kantor, *k-homogeneous groups*, Math. Z. **124** (1972), 261–265.

[77] _____, *Plane geometries associated with certain 2-transitive groups*, J. Algebra **37** (1975), 489–521.

[78] _____, *Homogeneous designs and geometric lattices*, J. Combin. Theory, Series A **38** (1985), 66–74.

[79] _____, *Flag-transitive planes*, in: Finite Geometries (Winnipeg, Can.,1984), ed. by C. A. Baker and L. M. Batten, Lecture Notes in Pure and Applied Math. **103**, Dekker, New York, 1985, 179–181.

[80] _____, *Primitive permutation groups of odd degree, and an application to finite projective planes*, J. Algebra **106** (1987), 15–45.

[81] _____, *2-transitive and flag-transitive designs*, in: Coding Theory, Design Theory, Group Theory, Proc. Marshall Hall Conf. (Burlington, VT, 1990), ed. by D. Jungnickel et al., J. Wiley, New York, 1993, 13–30.

[82] P. Kaski and P. R. J. Östergård, *Classification Algorithms for Codes and Designs*, Algorithms and Computation in Math. **15**, Springer, Berlin, Heidelberg, New York, 2006.

[83] P. Kaski, P. R. J. Östergård, and O. Pottonen, *The Steiner quadruple systems of order* 16, J. Combin. Theory, Series A **113** (2006), 1764–1770.

[84] P. B. Kleidman, *The finite flag-transitive linear spaces with an exceptional automorphism group*, in: Finite Geometries and Combinatorial Designs (Lincoln, NE, 1987), ed. by E. S. Kramer and S. S. Magliveras, Contemp. Math. **111**, Amer. Math. Soc., Providence, RI, 1990, 117–136.

[85] P. B. Kleidman and M. W. Liebeck, *The Subgroup Structure of the Finite Classical Groups*, London Math. Soc. Lecture Note Series **129**, Cambridge Univ. Press, Cambridge, 1990.

[86] F. Klein, *Das Erlanger Programm (1872). Vergleichende Betrachtungen über neuere geometrische Forschungen. Einleitung und Anmerkung von H. Wußing*, 3rd ed., Harri Deutsch, Frankfurt am Main, 1997.

[87] H. Kurzweil and B. Stellmacher, *The Theory of Finite Groups. An Introduction*, Universitext, Springer, Berlin, Heidelberg, New York, 2004.

[88] C. W. H. Lam, L. Thiel, and S. Swiercz, *The non-existence of finite projective planes of order* 10, Canad. J. Math. **41** (1989), 1117–1123.

[89] R. Laue, S. S. Magliveras, and G. B. Khosrovshahi, *t-Designs*, (book in preparation).

[90] M. W. Liebeck, *The affine permutation groups of rank three*, Proc. London Math. Soc. **54** (1987), 477–516.

[91] ———, *The classification of finite linear spaces with flag-transitive automorphism groups of affine type*, J. Combin. Theory, Series A **84** (1998), 196–235.

[92] D. Livingstone and A. Wagner, *Transitivity of finite permutation groups on unordered sets*, Math. Z. **90** (1965), 393–403.

[93] H. Lüneburg, *Die Suzukigruppen und ihre Geometrien*, Lecture Notes in Math. **10**, Springer, Berlin, Heidelberg, New York, 1965.

[94] ———, *Fahnenhomogene Quadrupelsysteme*, Math. Z. **89** (1965), 82–90.

[95] ———, *Some remarks concerning the Ree groups of type* $(^2G_2)$, J. Algebra **3** (1966), 256–259.

[96] ———, *Transitive Erweiterungen endlicher Permutationsgruppen*, Lecture Notes in Math. **84**, Springer, Berlin, Heidelberg, New York, 1969.

[97] ———, *Translation Planes*, Springer, Berlin, Heidelberg, New York, 1980.

[98] ———, *Ein einfacher Beweis für den Satz von Zsigmondy über primitive Primteiler von* $A^n - 1$, in: Geometries and Groups, ed. by M. Aigner and D. Jungnickel, Lecture Notes in Math. **983**, Springer, Berlin, Heidelberg, New York, 1981, 219–222.

[99] F. J. MacWilliams and N. J. A. Sloane, *The Theory of Error-Correcting Codes*, North-Holland, Amsterdam, New York, Oxford, 1977; 12. impression 2006.

[100] E. Maillet, *Sur les isomorphes holoédriques et transitifs des groupes symétriques ou alternés*, J. Math. Pures Appl. **1** (1895), 5–34.

[101] E. Mathieu, *Mémoire sur l'étude des fonctions de plusieurs quantitiés*, J. Math. Pures Appl. **6** (1861), 241–323.

[102] ———, *Sur la fonction cinq fois transitive de 24 quantités*, J. Math. Pures Appl. **18** (1873), 25–46.

[103] R. Mathon, *Constructions for cyclic Steiner 2-designs*, Ann. Discrete Math. **34** (1987), 353–362.

[104] E. H. Moore, *Tactical memoranda I-III*, Amer. J. Math. **38** (1896), 264–303.

[105] T. Nagell, *The diophantine equation $x^2 + 7 = 2^n$*, Nordisk Mat. Tidskr. **30** (1948), 62–64; Ark. Mat. **4** (1961), 185–187.

[106] T. G. Ostrom and A. Wagner, *On projective and affine planes with transitive collineation groups*, Math. Z. **71** (1959), 186–199.

[107] D. Pei, *Authentication Codes and Combinatorial Designs*, CRC Press, Boca Raton, 2006.

[108] G. Pickert, *Projektive Ebenen*, 2nd ed., Springer, Berlin, Heidelberg, New York, 1975.

[109] S. Ramanujan, *Collected papers of Srinivasa Ramanujan, ed. by G. H. Hardy, P. V. S. Aiyar and B. M. Wilson*, Amer. Math. Soc., Providence, RI, 2000.

[110] D. K. Ray-Chaudhuri and R. M. Wilson, *On t-designs*, Osaka J. Math. **12** (1975), 737–744.

[111] R. M. Robinson, *The structure of certain triple systems*, Math. Comput. **29** (1975), 223–241.

[112] M. Roitman, *On Zsigmondy primes*, Proc. Amer. Math. Soc. **125** (1997), 1913–1919.

[113] J. Saxl, *On finite linear spaces with almost simple flag-transitive automorphism groups*, J. Combin. Theory, Series A **100** (2002), 322–348.

[114] J. Siemons, *Orbits in finite incidence structures*, Geom. Dedicata **14** (1983), 87–94.

[115] J. Steiner, *Combinatorische Aufgabe*, J. Reine Angew. Math. **45** (1853), 181–182.

[116] D. R. Stinson, *Combinatorial Designs: Constructions and Analysis*, Springer, Berlin, Heidelberg, New York, 2004.

[117] _____, *Cryptography*, 3rd ed., CRC Press, Boca Raton, 2005.

[118] M. Suzuki, *On a class of doubly transitive groups*, Ann. Math. **75** (1962), 105–145.

[119] L. Teirlinck, *Non-trivial t-designs without repeated blocks exist for all t*, Discrete Math. **65** (1987), 301–311.

[120] J. Tits, *Sur les systèmes de Steiner associés aux trois "grands" groupes de Mathieu*, Rendic. Math. **23** (1964), 166–184.

[121] V. D. Tonchev, *Combinatorial Configurations: Designs, Codes, Graphs*, Longman, Harlow, 1988.

[122] H. N. Ward, *On Ree's series of simple groups*, Trans. Amer. Math. Soc. **121** (1966), 62–89.

[123] H. Wielandt, *Finite Permutation Groups*, Academic Press, New York, 1964.

[124] R. J. Wilson, *The early history of block designs*, Rend. Sem. Mat. Messina Ser. II **9** (2003), 267–276.

[125] E. Witt, *Die 5-fach transitiven Gruppen von Mathieu*, Abh. Math. Sem. Univ. Hamburg **12** (1938), 256–264.

[126] _____, *Über Steinersche Systeme*, Abh. Math. Sem. Univ. Hamburg **12** (1938), 265–275.

[127] K. Zsigmondy, *Zur Theorie der Potenzreste*, Monatsh. für Math. u. Phys. **3** (1892), 265–284.

Index

Frontiers in Mathematics

This series is designed to be a repository for up-to-date research results which have been prepared for a wider audience. Graduates and post-graduates as well as scientists will benefit from the latest developments at the research frontiers in mathematics and at the "frontiers" between mathematics and other fields like computer science, physics, biology, economics, finance, etc.

Advisory Board

Leonid Bunimovich (Atlanta), Benoît Perthame (Paris), Laurent Saloff-Coste (Rhodes Hall), Igor Shparlinski (Sydney), Wolfgang Sprössig (Freiberg), Cédric Villani (Lyon)

Printed in the United States
By Bookmasters